Insects' Habitats

Insects' Habitats

Edited by:

Laichattiwar Mukesh Anandrao

Delve
PUBLISHING

www.delvepublishing.com

Insects' Habitats

Laichattiwar Mukesh Anandrao

Delve Publishing
224 Shoreacres Road
Burlington, ON L7L 2H2
Canada
www.delvepublishing.com

Email: orders@arclereducation.com

ISBN: 978-1-77407-758-0 (Hardcover)

ABOUT THE EDITOR

Dr. Mukesh Laichattiwar, Ph. D (Institute of Agricultural Sciences, Banaras Hindu University, Varanasi, U.P.) is presently working as Assistant Professor at College of Agriculture, Parul University, Vadodara. Dr. Laichattiwar worked as Assistant Professor at college of Agriculture, Naigaon (Bz.), Nanded M.S. (2018-2019). He did his B.Sc (Ag.) From College of Agriculture, Ambajogai (Vasantrao Naik Marathwada Krishi Vidyapeeth, Parbhani, M.S.) and M.Sc (Ag.) from Institute of Agricultural Sciences, Banaras Hindu University,Varanasi, U.P., He was awarded UGC fellowship for his Ph. D research. He qualified ICAR-NET in 2013. Dr. Laichattiwar is a recipient of Best Poster presentation award in 2014. He Published 18 research article, 7 book chapters/proceedings, 5 popular articles and 15 abstract.

TABLE OF CONTENTS

LIST OF FIGURES

green plants, and apart from a few insects, such as spiders, there is hardly anything else inside the cave. After walking in a distance, one may note a fly or a moth on the wall, some fungi growing on a log or a different species' spider web

Figure 4.2: Some of the commonly known caves include the Time Machine in Daren Cilau which is the largest passage; Ogof Draenen which is the second -largest cave running approximately 70km in length; and Ogof Ffynnon Ddu which is the deepest cave in Britain, going around 308 meters deep

Figure 4.3: Troglofauna is a collective name used to refer to the creatures found and living within caves, and these creatures can undergo classification in accordance to how dependent they are to the environment inside the cave, and how much of their life cycle they spend in the cave

Figure 4.4: Above is a great illustration of the different zones that are found in the cave: the entrance zone, the twilight zone and the dark zone. The three zones are differentiated above

Figure 4.5: The troglofauna are very sensitive to small changes in their environment, especially because they are restricted in them for long periods. Floodwaters are considered one of the major threats to these species, due to the fact that thus, environmental change can lead to a reduction in oxygen and food availability, as well as space for them to move in the cave

Figure 4.6: Troglobite species are very unique and cannot complete their life cycle in any other environment other than the inside of the cave. This is so much that they have had to adapt morphologically to the environment inside the cave, for example, they have a lower metabolism, longer legs, and they are blind

Figure 4.7: Above is a picture of the tooth cave spider. It is classified as a troglobite because it lives a notably different lifestyle from its counterparts, the Amauropelma, mainly because it lives in in the darker recesses of limestone caves

Figure 4.8: With movement from outside into the cave, the nature of plant life changes, and this is correspondent to the habitat changes. There is a notable w the amount of light (decrease), and with an increase in the shelter, comes with a decrease in the wind, thus helping in the maintenance of humidity (in combination with the cave's damp air)

Figure 4.9: Conservation of these caves has proven to be a very important act. Caves are a rare and special environment, some being more so than others, and even some specific parts more than others. Cavers have the ability to accidentally damage potholes with fast -moving water and also active streams, which may have a major effect on the cave environment

Figure 4.10: Most insects within the cave are found nearer the surface where there is much more food. When going deeper into the cave the harder it gets to find more insects. Much deeper into a cave, there may be insect communities that are concentrated around a certain area. This area may have a source of food that the cave bacteria generates, such as flowstone

Figure 4.11: Springtails are small creatures that love moisture. They have no wings and

PREFACE

Insects are small animals with segmented bodies and with six or eight legs, some have exoskeletons, some have wings and others don't. A few examples we have mosquitoes, ladybirds, and ants. Insects live in almost every habitat on earth such as deserts, grasslands, forests, and also waters. Grasslands are one of the habitats that mainly consist of grass and shrubs and very few trees. They are homes for various insect species where insects like termites, ants, and butterflies live and depend on them for food.

Deserts, which are known to be arid and receive very little precipitation also host insects such as various beetle types and locusts. For example, locusts have hard skin as an adaptation to living in deserts. Also, insects such as stink bugs are found in deserts whereby they live underground and on rocks too.

Forests are very large areas with many tree species and other plants too. Forests host many insect species in them since there are different trees and most insects live there because they perform various roles such as carrying out pollination, some act as decomposers. Insects also live in forests as they have their food sources in the forest.

Waters such as lakes, large rivers, oceans also inhabit some of the world's insects such as the dragonfly, and water bugs. These insects are also referred to as aquatic insects as they spend their lives in water sources. Some insects may just have their live portion on water areas due to their life cycles where they may end up living in other areas such as dry land.

A large percentage of insects such as caterpillars, bees and small six -legged insects live on land. Some of them have exoskeletons that help in the prevention of water loss from the insects' body as an adaptation to living on land.

Tundra also supports different types of insects like bumblebees, deerflies, blowflies. Many of these insects feed on low vegetation such as moss and lichen. We also have weevils present in tundra areas that live and eat decaying vegetation.

There are also insects that live in polar areas such as bugs found in the Arctic Circle. Flies are known to be the most common bugs found in these cold areas.

From this volume, we find out the different habitats where different species of insects spend their entire lives on earth and some adaptations of the insects.

1

Where Do Insects Live

CONTENTS

1.1 INTRODUCTION

Figure 1.1: A cathedral mound created by termites (Isoptera).

Source: https://en.wikipedia.org/wiki/Insect

Insects are animal species that emerged about four hundred million years ago. Many of the insects live together in the same area while others are solitary. There are more than one million species of insects and they make up about two- thirds of the earth's species. Insects are well suited for the habitats they live in, to ensure their survival. Insects live in different habitats on the earth. They exist in different ecosystems including hot deserts and even cold areas of the world. Most insects are terrestrial and about 97% of them live on land and the rest live on other ecosystems. A variety of insect species spend most of their time on land such as butterflies, bees, and moths among others. Although many insects are considered or seen as pests, they play an important role in the ecosystem. Insects are important in population control, they also facilitate the process of decomposition of dead matter, they are important in the nutrient recycling process, they clear waste from other animals, they pollinate flowers and they can also provide food to animals and human beings alike.

Insects are members of Arthropoda. They have segmented bodies, have six jointed legs, and have hard exoskeletons. Some have wings depending on their stage of development. They are organisms that have close relations to insects such as scorpions, centipedes, and millipedes but they differ in morphological characteristics. Insects are some of the most versatile organisms. They live in the different environments of the earth some of these environments include trees, hives, burrows, water, rotten wood, loose barks, plant galls, flowers, woodpiles, aquatic areas, and more others.

Proturans are always found in moist habitats and this is usually in the humus and leaf mold of forests that are in temperate regions. Diplurans live in soil that is moist, leaf litter and humans. Springtails are among the species that are most abundant in soil. The silverfish are insects that are fast running and hide under stones or leave during the day and emerge after dark and they can be termed as nocturnal. Dragonflies and damselflies are predacious.

1.2 TREES AS HOME TO INSECTS

Insects perform various roles within forests or trees as herbivores, pollinators, as decomposers, carnivores, and as sources of food for other organisms. Most insects are considered harmful especially to plants or trees by humans. Insects that are commonly studied in forests are those that cause harm most of the time. They are classified into four categories such as defoliating insects which infest leaves; sucking insects which feed on sap; cone and seed insects which feed and live on the roots and stems of trees and finally boring insects, which feed and reproduce on trees.

Bark beetles are one of the most commonly known groups and consist of many species of which less than a dozen can lead to the death of trees. They are cylindrical and are often black or brown. The male and females bore into trees and form chambers for their eggs. After hatching of the eggs, the larvae bore away from the chambers and form channels as a result. A pupa is formed at the tunnel ends and later emerge as adults through the same holes bored in the tree barks. The different species of bark beetles attack specific trees and cause damage to their roots, stems, seeds, and even fruits. Diseases in plants can also be transmitted through the same beetles (Collins, 2012).

Figure 1.2: The bark beetle *Dendroctonusvalens.*

Source: https://www.globalwoodmarketsinfo.com/spruce-beetle-bc-omineca/

Figure 1.3: Bark beetles on a log pole pine at Yellowstone national park.

Source: https://oregonwild.org/about/blog/bugs-and-burning-logging-beetle-killed-trees-dont-stop-forest-fires

Weaver ants in largely forested areas for example the sub- Saharan Africa in trees. They construct structures that are complex using leaves and glue that is natural. They search trees to look for leaves that are suitable and attempt to bend them. When the leaf can bend, other ants hold on to the first ant in an attempt to assist it in pulling the leaf back into a shape that is pocket like for building nests. Adult insects use silk produced by larvae as glue to seal the leaves shut. As more and more larvae leaves are pulled along, a lump of green leaves that are fresh and lined with a white silk mat is formed in the end. Larvae producing silk is not used to spin cocoons but is rather used in the creation of structures which offer protection from predators. Damage can be caused to the nest by strong winds and leaves that are wilted. This damage means that the worker ants have to continuously keep constructing many more in the same way. Weaver ants as observed dominate an entire canopy of a single tree and do these before moving to the next tree (Kalman & Crossingham, 2006).

Figure 1.4: Weaver ants working together to build a nest.

Source: https://www.entoblog.com/about-the-weaver-ant/

The eastern tent caterpillar moth is named after the nest it forms which look like tents. Their eggs hatch early in the spring at a point where the temperature in the air is still low. Up to 300 caterpillars that are emerging gather on treetops to spin a silk tent. In the day time, the temperatures are high in comparison to their environment and this raises their temperatures improving metabolism. The tents comprise thin layers of silk with the use of branches from the tree it is inhabiting as attachment points. It is expanded three or four times each day by all larval stages except its last stage in which it saves its silk to spin a cocoon. At points where the branches stick out, entry and exit points are maintained from the tent. In the event they need to feed, they form bands together during the journeys to and from the sites of feeding using visual and chemical cues.

Figure 1.5: Communal tent of the eastern tent caterpillar.

Source: https://en.wikipedia.org/wiki/Eastern_tent_caterpillar

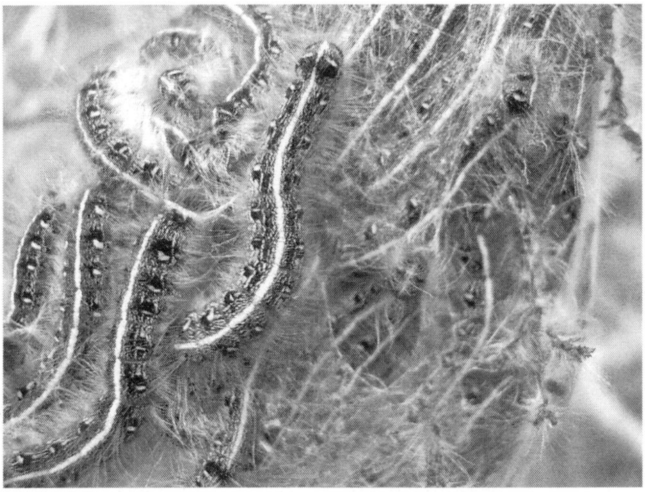

Figure 1.6: The tent after hatching of the caterpillar.

Source: https://en.wikipedia.org/wiki/Eastern_tent_caterpillar

Aphids can also infest trees especially the leaves or shoots and can cause damage to the heath of plants and trees. They can be found on trees such as spruce, conifers, fir, maples, elm, and pine among others. Some include the Norway maple aphid, birch aphids, pine aphids, balsam wooly aphids and spruce gall aphid among others. Most of them are named after the tree they are found on. They feed on them and the signs of severe feeding of aphids are leaves that are twisted and curled, yellow leaves, dead shoots that are stunted and growth that is stunted. Aphids are pear- shaped and insects that have soft bodies. Mature aphids mostly lack wings and they can sometimes have wings. Aphids can be specific to trees or shrubs and therefore do not attack or live on any other tree type. They can blend in as they take the green color of plants many times. Throughout winter the eggs of the aphids remain of the leaves or barks of trees and in spring they turn in to larvae which also develop at the sell sap expense and they molt after insect reproduction. Aphids that are wingless leave a life that is sedentary and they can sometimes spend their entire life at the same place. They are parasites, they reproduce in the same habitat and cause harm to vegetative matter. On the other hand, aphids with wings can migrate to new locations and ensure continuity of their species once they have exhausted food resources (Hopkins, 2009).

Figure 1.7: An image of giant conifer aphids on a conifer tree.

Source: https://www.whatsthatbug.com/wpcontent/uploads/2015/07/giant_conifer_aphids_irja.jpg

The nymphs of the meadow spittlebug live in a mass of their spit which is frothy and is a product that comes from their anuses. Shortly after they hatch in spring, they go looking for spots in meadows that are shaded by sunlight and winds. After feeding on juices from plants, they start producing spittle. They first begin by secretion of anal fluids that stick to and somehow submerges the body blow into the liquid to form bubbles. The nymphs molt and metamorphose into adults in the bubble bath itself and they leave from time to time for feeding purposes. During the final molt, the production of spittle stops and the nest is left to dry out.

Figure 1.8: The nymph of the spittlebug in a mass of bubble fluid secretion.

Source.www.insectsofalberta.com/spittlebug.htm

Case moths get their names from their habit of leaving cases at the tree bases. They extend their heads from cases to feed on leaves. They defoliate trees particularly those that are evergreen usually at branch ends. The black turpentine beetle bores into pines that are damaged or stressed in the south. They attack the phloem layer and may lead to the death of the tree. Their young ones boreholes in the bark and may cause blue stain fungus (Hopkins, 2009).

Figure 1.9: The larvae of the case moth extending its head from the case for feeding.

Source: https://en.wikipedia.org/wiki/Metura_elongatus

The locust borers have their larvae tunneling into and through the trunks and branches of trees that have been infested and this leads to them being damaged physically and making them susceptible to breakage by the wind. Their attacks that are repeated often result in reduced reproduction. Their larvae are hatched bore into the inner back and construct cells for hibernation. The Nantucket pine moth causes damage that is severe to shoots and buds and shoots of pine seedlings and small saplings. Larvae that are newly hatched feed on new growth for a short period before boring into the shoot.

Figure 1.10: The locust borer.

Source: mobugs.blogspot.com/2010/09/locust-borerh.html

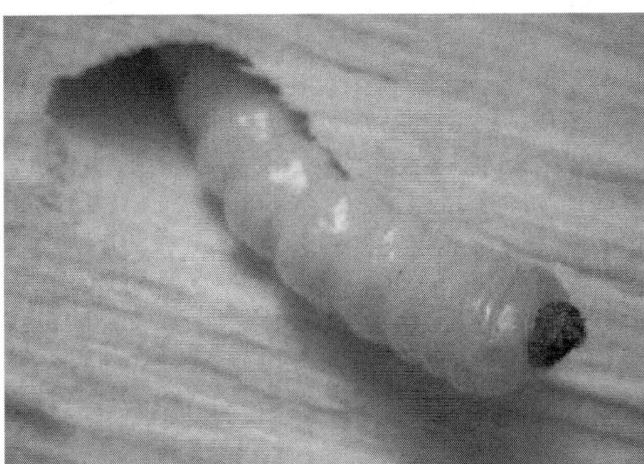

Figure 1.11: The larvae of the locust borer.

Source: bugspray.com/purple-robe-tree-borer-control.html

The pine webworm selects seedlings that are young but can still attack saplings and trees that are large. Their feeding habits can lead to needle loss but do not cause the death of trees.

Figure 1.12: Larvae of the pine webworm moth.

Source: https://en.wikipedia.org/wiki/Pococera_robustella

Figure 1.13: Damage caused to pine trees by the pine webworm moth.

Source: https://en.wikipedia.org/wiki/Pococera_robustella

Other insects that live on trees include the ashplant bugs, birch leaf miner, cicadas, emerald ash borer, the flea weevil, galls; different species of month, honey locust plant bugs, the gypsy moth; beetle species, lace bugs and sawflies (Hopkins, 2009).

1.3 HIVES AS HOME TO INSECTS

Other insects use different kinds of materials while building their homes. Honey bees build elaborate homes and referred to as hives and they are built from wax organized in hexagons which areused for the storage of honey.

Bees collect nectar from flowers and store them in their stomachs before returning to their hives. It is then evaporated into a form that is thick and is the honey we know.

They have different bee classes carrying out different functions in their colonies. The worker bees develop glands that are special that allow the conversion of honey into wax. The wax then oozes through pores onto their abdomen and the bees then collect and chew it making it into a substance that is pliable which they then use in the construction of their habitat. The beehives are built in the branches of trees, hollow spaces in trees or rock crevices (Aloian, 2018).

Figure 1.14: A hive of Morven bees.

Source: https://morvenbees.wordpress.com/2012/10/30/bee-publicity-and-intro-to-winterizing-beehives/

Carpenter bees create nests in wooden objects in houses or outside. They are large but at the same time they are harmless and do not sting like the honey o solitary bees. They are considered harmful when they inhabit wooden structures, walls, or roofs. They bore into wood rather than building nests or hives and lay their eggs at each end.

Figure 1.15: A carpenter bee attacking a piece of wood to make a tunnel for nesting.

Source: https://www.thriftyfun.com/getting-rid-of-carpenter-bees-1.html

The small hive beetle has an African origin and their larvae damage capping and coms and it also ferments honey. They defecate on the honey discoloring it these causes the honey to develop a characteristic odor. The varroa mite lives in bees that are developing and they kill and deform them and they also attach themselves to bees in their adult stage. Adult wax moths also enter hives to lay their eggs and they also go for the pollen if there is any in hives (Aloian, 2018).

Figure 1.16: A hive infested with the larva of small hive beetles.

Source: https://en.wikipedia.org/wiki/Small_hive_beetle

Hornets and wasps also make nests but in contrast, they are made with a substance that has similar characteristics to paper and is created from chewing of wood. Unlike bees, they do not ingest nectar for making honey and wax.

The nest of wasps varies depending on the species. Yellow jackets are among the common wasp species and they create their habitats by chewing fiber from wood. Wasps round the resulting substance that is pulpy from a fiber mix and saliva into combs that are stacked on top of each other. Additional pulp covers the nest. Wasps create a nest in areas that are secluded most of the time and this includes buildings, trees and hedges. Paper wasps are the species of wasps that commonly build their nests in buildings, under eaves and beams. Such nests lack an additional covering of pulp like others and the actual comb becomes visible (Kalman, 2009).

Figure 1.17: The nest of the yellow jacket wasp.

Source: www.ecotechli.com/about-the-critters/nests/

Figure 1.18: A paper wasp nest.

Source: https://www.agric.wa.gov.au/pests/paper-wasps-wa-and-their-control

1.4 BURROWS AS HOMES TO SOME INSECTS

Some insects inhabit underground ecosystems. Insects such as termites create mounds above and also below the ground, creating tunnel systems that are extensive connecting their home. These mounds reach up to seventeen feet high and can weight up to ten tones. These habitats have been studied by scientists and it is amazing how such small insects can move large amounts of soil. Termites live in colonies of up to millions of members each having their own responsibilities in mound building. Scientists have taken part in experiments to see how extensive the tunnels were and have since then established amazing findings. The tunnels were large and they were all connected to the mound at top and around six feet below the surface (Hopkins, 2009).

Figure 1.19: Termite mounds in Namibia.

Source: https://en.wikipedia.org/wiki/Termite

The mason bee burrows a few centimeters below the soil that is deep upon finding suitable nesting sites. The nest usually has three layers.It builds the outer layer using flower petals it gets from neighboring plants and smears them inside soil which has some moisture. It lets the soil dry and collects more petals to line its nest and eventually creates a sandwich of between soil petals. Once the nest has food, the female lays its eggs. Before leaving the nest, the bee seals its entrance by folding the petals. The petals are useful in the retention of moisture for larval development at the same time preventing the flooding of this habitat. Trapped air facilitates floating in the event of a flood while the mud that is hardened prevents the larvae from predator attacks (Hopkins, 2009).

Figure 1.20: Mason bees building their underground nests.

Source: https://www.habitataid.co.uk/blog/garden-plants-for-butterflies-and-bees/

The leafcutter bee nests in wood cavities and also burrows the soil. Using its jaws as scissors. It cuts leaves in oval or circular discs and uses them to line its nest. This partition forms the base of the next cell and the process is done repeatedly until it is complete. A single cell takes up to ten hours of construction inclusive of breaking for meals. This includes twenty one trips of collecting leaves especially from the rose plant and it takes eighteen trips to collect food for stocking.

Figure 1.21: The leafcutter bee underground nest.

Source: https://content.yardmap.org/learn/bare-earth-for-native-pollinators/

Figure 1.22: The leafcutter bee cutting pieces of leaves for building a nest.

Source: realworldgardener.blogspot.com/2015/06/green-colours-but-not-on-potatoes.html

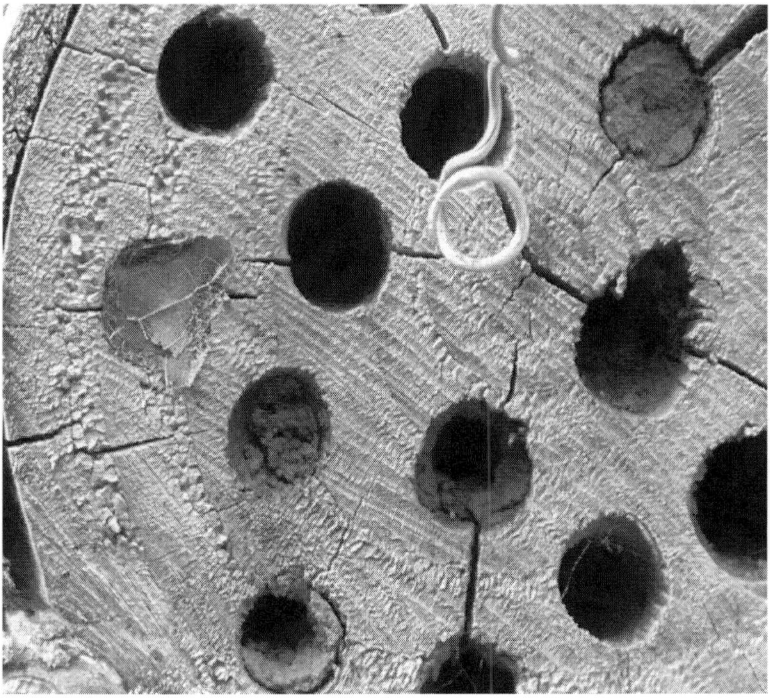

Figure 1.23: Leafcutter bee nest in wood cavities.

Source: abugblog.blogspot.com/2011/07/leaf-cutter-bee-antics.html

The yellow meadow ant builds mound nests which are conspicuous on pastures that are not disturbed and moorlands across Europe. Worker ants bring soil beneath the ground to the surface making a heap that is porous. Each heap can reach up to 0.5 meters in height and a diameter of up to 1.5 meters can exist for more than one century. The turf that is overlying functions to reinforce the mound. The mounds are built in open lands most of the time and can openly face the sun, often encircling boulders for support and trapping of heat. Ants in meadows spend most of their time below the ground. These conditions favor aphids and their honeydew secretions are fed on by the ants. The ants also spend the winter seasons in hibernation mostly in galleries that are below ground. During the late summers, the hordes of winged queens come out from the nests to create new colonies (Hopkins, 2009).

Figure 1.24: Yellow meadow ants.

Source: northwestnorfolknaturalists.blogspot.com/2014/04/redgate-hill.html

Mole crickets are common in urban parks and gardens that are well watered. The males use their forelegs to dig burrows that are specially constructed which act as amplifying horns. The males that are flightless can be heard in the evening during months that are warm and can be very loud, make continuous calls using their wings which are modified. The call facilitates the process of mating. The sand gropers are large burrowing insects from the Orthoptera family and are commonly found in Western Australia. They are sometimes considered as pests and both of the sexes do not possess wings and they usually come out from the ground. The Cooloola monsters are an Australian family that is endemic and are commonly found in sandy coastal areas especially in different parts of Queensland. They are also burrowing insects that cannot fly. They are called monsters because of their bodies which are large and robust and their strong front legs which have claws. Only a portion of their species have been researched on and described so far as they are of rare species.

Figure 1.25: Mole cricket with fossorial legs for burrowing.

Source: https://www.bugs.com/bugs_database/lawn_pests/molecrickets.html

Digger bees are also called mining bees and are bees that are solitary and can create small burrows in the soil. In most cases, several queen digger bees tunnel their burrows close together and are important as pollinators. They collect pollen and nectar from their young ones. They are similar sizes to the honey bee and they can also be smaller in size and therefore they vary in their appearance. Others are metallic green and some are stripped. They can hover around a singer site or area. They rarely sting and therefore should not be a cause for alarm.

Figure 1.26: The digger bee coming from its underground nest.

Source: allthedirtongardening.blogspot.com/2012_03_01_archive.html

Cicada killer wasps can grow up to more than one and a half inches and have black abdomens with yellow or orange stripes. The female digs up to four feet with numerous tunnels that are branching. It burrows into the soil that is loose including sand on which people play in and even gardens with vegetables that are tilled. It hints cicadas exclusively. They are paralyzed by females and then stuffs them into an underground cell for brooding to feed their young ones at their larval stage. They are insects that are solitary and they may build their nests close to others. The males over anything but they do not sting but the females on the other hand are large stingers but are not really aggressive and they also don't protect their nests all the time like other species of Hymenoptera (Hopkins, 2009).

Figure 1.27: A ground hole of the cicada killer wasp.

Source: https://myhandsinfathersgarden.wordpress.com/2012/08/04/cicada-killers/

They Cydnidae group of insects are referred to as the burrowing bugs and are family of small bugs that are black in color and can be confused with beetles. They tend to live in the soil and they feed on roots and other plant material and the climb on plants for feeding purposes. Nymphs or adults can be found in shallow soil like gardens and they try to burrow as soon as they are exposed. The adults can be found in open spaces and sometimes on plant tips. They are visible since their color contrasts with that of plants.

Figure 1.28: Cydnidae beetle also referred to as the burrowing insects.

Source: https://www.flickr.com/photos/carlos_mancilla/28970474235

Soil insects have variations in their morphological characteristics for example some of them have eyes with reduced sizes and their wings are also protected by front wings that are hard and reduced. Most ground insects are flightless and therefore use jumping as their primary form of movement.in these insects, their front legs can also be modified for digging purposes and their legs are Fossorial.

There are seasonal changes in the distribution of subterranean insects. At greater soil depths, the temperatures attract insects during winter periods. The water level in soil is an important factor that controls distributions that are either be vertical or horizontal as some of them will seek sites that are drier for their pupa. Ants and termites have their nests located in areas that are dry and the point of entry is also elevated above the ground to prevent the process of flooding. The location and design of the nests have its importance in relation to the temperature and humidity regulation because ants and termites cannot ventilate their nests using fans. Most insects that live in the soil get their food from consuming large volumes of soil which contains vegetables that are dead or decaying and also associated with microbes. Detritivores include insects such as the collembola, beetle larvae and some termite species (Hopkins, 2009).

1.5 WATER AS HOME TO SOME INSECTS

Other than trees living in trees, the air and on land, some of the insects live in water. Aquatic insects live some part of the life cycle in water and they feed in the same way as other insects. Some insects that dive such as the predatory diving beetles can hunt for food while in the water where those insects that live on the surface cannot. The one problem insects that live in water must overcome is gaining access to oxygen while in water. Insects draw air into their bodies using their spiracles. They have several adaptations that enhance their survival in water (Kalman & Crossingham, 2006).

The larvae and nymphs of mayflies, dragonflies and stoneflies have trachea but in their larval stage, it is connected to the gills. A few insects such as water scorpions and the larvae of mosquitoes have tubes that they use for breathing without leaving the water. Aquatic and semiaquatic insect species include the springtails, dragonflies, mayflies, stoneflies, fish flies, lacewings, beetles, moth, caddisflies, scorpionflies, flies, ants, water striders, and wasps.

The water strides fall under the Gerridae family. They live on surfaces where there are waters that are calm and fresh an exception are the Asclepios, Halobates and Stenobates which will leave in marine areas. The Gerridae will prefer areas that have a large number of insects or zooplanktons and similarly those that have plants or rocks for them to have places to oviposit their eggs. Water strides will prefer areas with room temperatures and temperatures that are lower than twenty-two degrees Celsius are not favorable since the development of their progeny is dependent on temperature and therefore the surrounding environments which are cool slow the rate of development. They are found predominantly in Europe, Canada, the United States, South Africa, South America, Australia, China and Malaysia. They are aquatic predators and they consume insects that fall on water surfaces and also spiders (Kalman & Crossingham, 2006).

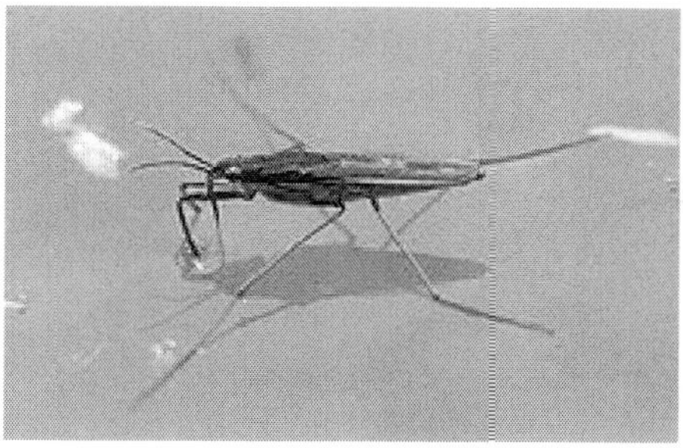

Figure 1.29: A water strider using surface tension to float on water.

Source: https://en.wikipedia.org/wiki/Gerridae

Adult fish flies are usually found among vegetation which is within different stream types which are required by the larvae. They are attracted to lights sometimes during the night. They do not live for a long time in their adult stage. They mate, lay eggs and then they die soon afterward. The different species will have a preference for different aquatic habitats. Some of their larvae will prefer waters that are littered, including stream edges while other larvae will prefer clear waters, some that flow faster and those that are oxygenated better.

May flies are distributed worldwide in habitats with freshwater but are absent in Antarctica, they are also absent in oceanic islands. Dispersal of the females is by wind and their eggs can be transferred by adhesion through the legs of water birds. Their main species have preferences in terms of where they live. Warm waters favor the Baetidae; Heptageniidae live beneath rocks and prefer waters that are fast- flowing and also large; the Epheneridae make burrows in sandy lakes or the beds of rivers. The nymphs of mayflies mostly live in streams under rocks, in vegetation that is decaying or in sediments. A few of the species live in lakes and this is among those that are prolific. They are sometimes found in muddy areas at the bottom of ponds. They are environments that are low in oxygen concentration ,for example, the *Ephemeravulgata* burrows and their gills which are filamentous act as organs for respiration and is used for the exchange of gases (Del-Claro & Guillermo, 2019).

Figure 1.30: The Tisza blooming, a tourist attraction in Hungary.

Source: https://hungarytoday.hu/tisza-mayfly-and-tisza-blooming-become-hungaricums-as-national-values/

Stoneflies can be of many types and live in a variety of habitats. When they are in their nymph stage they live in streams that are flowing, rivers and brisk springs. The stoneflies creep under rocks and objects that are submerged in water. Some of the mayfly species of insects prefer lakes, ponds and other quiet waters. Adult mayflies fly poorly and therefore they stay close to the water and in places that are shady such as beneath leaves or bridges. The species can also come out at night since they are attracted by artificial lights (Del-Claro & Guillermo, 2019).

Figure 1.31: A stonefly on a rock underwater.

Source: https://nature.mdc.mo.gov/discover-nature/field-guide/stoneflies

Springtails are semi-aquatic. They live in soil especially that is mixed with compost, litter made up of leaves and also organic mulches. They can also be found beneath barks or on wood that is decaying. They consume plant material that is decaying, fungi, mold and algae. In relation to aquatic habitats, they are found on the surfaces of water that is stagnant, swimming pools and also flower beds. They love environments that are moist and this can also be provided by mushroom and greenhouses. They require environments that are moist for their development. They search for damp areas if their environments become dry. When they enter houses, they crawl and search for moist areas such as sinks, washbasins and also tubs. They may also be found around floor drains and basements with moisture.

The caddis fly is a winged insect species that live in small moving water habitats such as freshwater sources around the world especially North America and Europe. Their larvae build a special kind of casing to them from their predators and also evade them. The larvae create thick cases that are in tubular form using twigs and roots of aquatic plants and embed in the shells of snails. These raw materials are cemented using the silk it spins to form a hard covering. The larvae drag this covering around with their bodies protruding out. This home which is now portable becomes safe for them and also assists the larvae to survive where there are shortages in the supply of oxygen in their development to pupae, the larvae plug the ends of the case and use the small holes to breathe (Rufer, 2015).

Figure 1.32: The larvae of the caddis fly in a case underwater.

Source: https://lifeinfreshwater.net/caddisfly-larvae-trichoptera/

1.6 LOGS OF WOOD AS HOMES TO INSECTS

There are two categories of insects that inhabit wood. There insects that will live on healthy trees and are referred to as primary invaders while those that live on stressed wood are referred to as secondary invaders. Most secondary invaders are commonly viewed as being invasive or a nuisance but both of them can take a horrible toll on trees. Destroying the wood inhabited by such insects doesn't stop the harm they bring and also their spread. A variety of insects will attack dead or trees that are dying. The activities of such insects ensure the breaking down of wood and their recycling. The most common insects found in wood are the beetles. They include the round- headed wood borers, the flat -headed borers and also the powder post beetles. The first two types, have legless larvae and can be identified in the process of splitting wood and piles of sawdust can also be observed in near wood infested with some of the beetles. Most of these insects are harmless except for wood. Carpenter ants and termites can also infest firewood (Hopkins, 2009).

The twig girdler lays their eggs inside the wood where they develop without being disturbed. The twig girdler, which has a brown-grey color chews grooves which have a v shape all around twigs and branches that have fallen. Their larvae have a difficult time living in wood with sap that is healthy and preventing their presence can be done by clearing any branches that fall from an area. The snout of the wood- boring weevil is made from chewing trees so that they can cause damage which is extensive when they attack the trunk of trees and the bases of woody plants that are ornamental. Their presence can be recognized by identifying their larvae which do not have legs and have a cream color and are found at the bark of trees.

Figure 1.33: The twig girdler on a pecan twig.

Source: https://en.wikipedia.org/wiki/Oncideres_cingulata

The larvae of the carpenter worm moths build tunnels in the trunks of oaks, the elm and the cottonwood trees. They invade the willow, pear and also cherry trees. As they develop into moths, their larvae move in and out of the tree trunks leading to damage that is significant still causing damage that is also significant. Their presence can be identified by piles that have a resemblance to sawdust in tree cracks and crevices.

More than six hundred species of bark beetles live in trees in the United States and Canada. They have the tendency of attacking different species of trees. The cedar bark and cypress beetle attack cypress trees, false cypress, junipers and redwood trees. The southern pine beetles attack pine trees that are stressed, week and injured. The beetles, which are small, red-brown or black tunnel beneath the bark of the tree and go into the wood. These activities turn the needles of pinewood into a red-brown color from their original green color within a one to two-month period after they have attacked the pine tree. Their spread can be controlled by the removal of the tree that is infested.

The seed bug can also be referred to as a true bug or the western conifer seed bug. It is pointed on both sides and it can grow up to one inch. The bug spends its summer in pine trees and it feeds on the tree sap. Their activities may cause the premature falling of pine cones and their habits do not cause damage to healthy trees (Aloian, 2018).

Winged termites leave on the cellulose of wood and they create multiple tunnels as they chew through the wood. The wood will look fine on the surface, but the wood crumbles if prodded using construction tools. Two of the most common types are the dry wood and the subterranean termites. Termites will invade wood that is close to the soil. They can also be found in basements with some moisture and most of the time they are found in the foundation of joists and structural lumber. Like their names, dry wood termites consume wood that is dry found in structural lumber, fence posts and the limbs of trees that are dead. They do not build underground and enter the homes of people that are above ground.

Powder post beetles are beetles with a brown color, with a long body frame and also possess small heads. They are less than one inch but this depends on the species. They lay their eggs on wood tops or beneath wood surfaces. The eggs then hatch to form larvae which feed on the wood and create tunnels to the wood surface and this can last for five years. They consume woods that can either be hard or soft and the type of wood will depend on the tree species. This damage can also be observed on floors, planks, paneling, and the frames of doors; furniture and also molding (Aloian, 2018).

Figure 1.34: The powder post beetle.

Source: https://smarterpestcontrol.com/pest-id/crawling-insects/powderpost-beetles/

Carpenter ants don't consume wood but they burrow through them to construct their nests. Morphologically they have a variety of colors depending on the species type they can either be black, yellow or brown. Like termites, some of them possess wings equal in length. Their fore wings are longer than their wings at the back. The ants have the perfect points of entry, especially when the wood is rotting. The ants can be found near wood in basements or near wood that is rotting around window sills or pipes that are leaking. They can be identified by the presence of wood shavings around areas that have been infested by the carpenter ants.

Figure 1.35: Holes in a tree from carpenter ants.

Source: https://en.wikipedia.org/wiki/Carpenter_ant

Wood borers feed on trees, acorns and nuts. Their activities lead to the creation of holes in the wood when they move out while in larvae form, but structurally they do not cause damage to the wood they infest. Wood -boring insects include the metallic wood borer, long -horned beetles and the wood wasps and their appearance will differ depending on their species. Wood borers will be found in log homes, the wood used in making structures and firewood (Aloian, 2018).

1.7 INSECTS OF LITTER, DEBRIS AND SOIL

Litter includes fallen vegetation, materials such as leaves, twigs, wood, fruit and flowers in various forms of decay. Some processes lead to the incorporation of such matter in the soil for example microorganisms such as bacteria and fungi. The acari, termites, ants and many beetles are important insects in litter and humus. The immature stages of many insects such as beetles, flies and moths can be found in both litter and soil. In forests found in Australia, litter with eucalyptus leaves are consumed by larvae of many oecophorid moths and certain chrysomelid leaf beetles.It is also composed of non-insect hexapods such as Collembola, Protura and Diplura and some insects which are wingless such as Archaeognatha and Zygentoma. In terrestrial litter, there exists insects of the ordersOrthoptera, Blattodea and Orthoptera as along withZoraptera, Embiidina and Grylloblattodea. Soils that are waterlogged can also have aquatic insects living in them (Collins, 2012).

The tranformation of recently fallen litter to decomposed matter is a gradual process. Insects will live in the different soil layers and have different morphological characteristics for example ,Onychurus from the Collembola family is found in deep soil layers, reduced limbs, blind and it lacks furcula. e Hypogastrura, found in intermediate soil has simple eyes, appendages which are short in contrast Collembola which live among litter which is superficial have large eyes, appendages that are long and a furcula that is elongated and is more than half the length of the body.

Field insects hide in leaf litter and they can also build burrows in the soil which they use to store food for later consumption. Other crickets include the young cave crickets. Termites are in the process of decomposition for wood that is decaying and can be found on the floor of forests. Termites live in various locations that are dry but also moist especially beneath wood and lifer that creates a moist environment. They convert the cellulose in dead decomposing matter with the assistance of microorganisms present in their intestinal tract which secrete enzymes used to digest the cellulose.

Earwigs are from the order Dermaptera and they are insects that are nocturnal and can be found in the vegetative litter. They are insects with wings that are leathery and on their abdomen they have pincers which they use to protect themselves. They feed on plant matter that is found in plant litter. Some of them feed on algae, fungal spores, algae and lichen and moss. They are consumed by other insects such as beetles, spiders, centipedes, birds and also lizards (Collins, 2012).

Figure 1.36: The earwig in a pile of leaf litter.

https://www.smithsonianmag.com/science-nature/meet-ecologist-who-wants-unleash-wild-backyard-180974372/

Thrips are found in leaf litter and have mouthparts for sucking purposes and they feed on fungal spores, hyphae and they sometimes feed on other arthropods that are small in size. They can either be winged or wingless. They are pests and are sometimes found on flower blossoms. Booklice consume algae, lichen, fungal spores and Mycella. They can either be found indoors or outdoors. Those that are found outdoors have wings.sixty species of the booklice are found in leaf litter.

True bugs are from the order Heteroptera and primarily feed on plants. They are found in leaf litter most of the time for example the stink bugs which produce repugnant smells that protect them from their predators. Others inhabit such areas during the winter period. True bugs that are present in leaf litter include the jumping ground bugs which are activated in the event of a disturbance; the lace bugs feed on plants but at the same time they are still found in leaf litter similar to pirate bugs which are minute and feed on arthropods which are small; seed bugs feed on seeds and flat bugs consume the sap of fungi (Kalman, 2009).

Figure 1.37: An adult female stink bug.

Source: https://njaes.rutgers.edu/stink-bug/identify.php

Beetles are from the order coleopteran. Beetles are diverse and can be found in different terrestrial ecosystems including leaf litter. They include the ground beetle found in North America. The rove beetle which plays an important role in the litter ecosystem as the feed on mites, collembola or insect larvae similarly they consume organic matter that is decomposing and also fungi. The featherwing beetles are also found in leaf litter and they feed on fungal hyphae and spores. Short winged mold beetles found in leaf litter and there are over 700 insects that are known. They feed on worms, collembola, insect larvae, mites and symphylans. Other beetles include the click beetles, the dung beetles and the carrion beetles. Weevils are largely known as plant and seed feeders but they can be found in leaf litter and they include the nut and acorn weevils (Kalman, 2009).

Figure 1.38: The short -winged mold beetle.

Source: https://en.wikipedia.org/wiki/Meloe_angusticollis

True flies belong to the order Diptera and they are found in different habitats including leaf litter. Adults found here lay eggs that then emerge as pupa or adults. Some of them are also parasites and are called humpbacked flies. Despite sometimes being a nuisance, their larvae can be important in the process of decomposition. Many species of wasps and bees sometimes build homes in the ground most of them are solitary. The layer of litter which is thick provides a layer of protection to insect especially, in the winter season. Wasps are important predators in relation to insects and the same time bees perform ecological functions such as pollination. Ants live in colonies that are organized but they can also be found in leaf litter (Kalman, 2009).

1.8 INSECTS OF DECAYED WOOD

Tree death can be a result of the activities of insects as they play a role in the distribution of fungi that is pathogenic. Different tree habitats have different insects but those inhabiting decayed wood are from three orders they include the Coleoptera, Diptera and Hymenoptera. Secondary orders come from the Lepidoptera and Heteroptera. The insects can also be classified according to systematics, their dependence, decomposition stages; the areas of the decaying tree where they are found and their nutrient consumption.

Species found in dying fresh dead wood include *Anthaxia candens, Acanthocinus aedilis, Ips typographus, Lamprodila festiva* and *Coraebus fasciatus* . These insect species feed on the sap under the back of the wood. The female *Coraebus fasciatus* lays its eggs on the tip of healthy twigs, their larva bore holes to the base of the twigs and they are circular leading to the death of the wood.

Figure 1.39: *Coraebus fasciatus* , the metallic beetle.

Source: https://www.forestryimages.org/browse/detail.cfm?imgnum=1433001

Insects can also feed and be found on the woody stems and they include species such as *Coleoptera cerambycidae, Hymenoptera formicidae and Hymenoptera apidae.* Other insects also feed on wood that is infested with fungus. Others will also be found within the cavities of dead or decaying wood especially from Coleoptera such as *Procraerus tibialis, Phloeophagus lignarius, Ampedus cadinalis and Elater ferrugineus.* Other insects also live in stumps that are rotten and also roots also from the coleopteran family. They include *Lucanus cervus* and *Dorcus parallelopipedus.* Predators and parasites also exist in such an ecosystem.

Dead wood insects play their role in the ecological system as decomposers. They also facilitate nutrient release from dead organic matter, they also provide nutrients required for the growth of plants. For example, the *Pine* species of insects help in the wood decomposition of the pine tree, the oak tree can also be decomposed by the *Querces* species. They also play the role of predators, habitat builders and pollinators.

The ambrosia beetles have an association with deadwood and the ambrosia fungus. In their adult stage, they eat into wood creating tunnels. As they do this they spread fungus which facilitates the decay of wood. As the wood breaks down it also becomes more nutritious. The adults and the larvae feed on the wood. Some insects are also attracted by forest or wood that have recently been burnt for example the *Henoticus serratus* which is commonly found in Europe. It spreads the ascomycete fungus and it is a source of food for other insects especially those that are pyrophilus for example, species of the jewel beetles and some flies (Kalman, 2009).

Figure 1.40: *Xylosandrus crassiusculus* split open, with pupae and black fungus.

Source: https://en.wikipedia.org/wiki/Ambrosia_beetle

Fallen or rotten timber provides a resource that is valuable especially for those that are detritivorous if they can overcome the problem of inhabiting a cellulose- rich substrate. Termites are some of the insects which are able to and survive on this diet type. Insects such as cockroaches are also able to produce the cellulose enzyme which allows them to digest cellulose (Kalman, 2009).

1.9 PLANT GALLS AS HOME TO SOME INSECTS

Galls are abnormal growths on plants caused by insects especially those that are immature and they can also be referred to as plant tumors. Most galls on plants are produced by midges, mites, and wasps. These organisms produce galls and then use them to make food and shelter provisions for their own gain. The galls they form lead to plant deformities such as leaves that are rolled which pocket like structures. These structures can also look like it is not made out of the plant tissue that is infested and seem highly organized. Plant mites are not insects but also lead to the production of galls (Aloian, 2018).

Another cause of plant galls are the aphids and plant lice. Both of them have mouthparts for sucking. The galls that they produce are complex. They spent winter on the barks and they hatch out to form galls during the spring period by attacking plant buds. They are the most common type of galls produced by plant lice include the hackberry nipple and the blister gall. During the spring, the large nipple gall forms beneath the leaves and the blister galls appear on top of the leaf when the insects start feeding. During the fall seasons they can enter into homes and become a nuisance to people.

The adelgids have similar features to the aphids. Some of the galls include the eastern spruce gall and the cooley spruce gall. They are commonly found on conifers. The eastern spruce gall is commonly found in Norway. They cause weakness to the stems leading to their breakage. The cooly spruce gall is commonly found on the blue and white spruce. Damaged caused to the trees can be rare unless the trees are young or they have multiple galls. After gall formation, the adelgids fly to the Douglas fir in the late summers to lay their eggs. They produce needles that are yellow and twisted in relation to the Douglas fir. During the fall season, the go back to infesting the spruce since they have wings and they also produce offsprings that live there during the winter period. The female species from this population lay eggs in spring which also produce galls (Aloian, 2018).

Gall midges are tiny flies that have delicate bodies. They produce galls in their larval stage. They spend their winter in their galls and emerge during the spring period. Some of the gall producing midges which are common include the willow pine cone gall, the gouty vein gall on maple trees and the filbert gall. They also infest other plants such as wheat, rice, soybean and cauliflower among others.

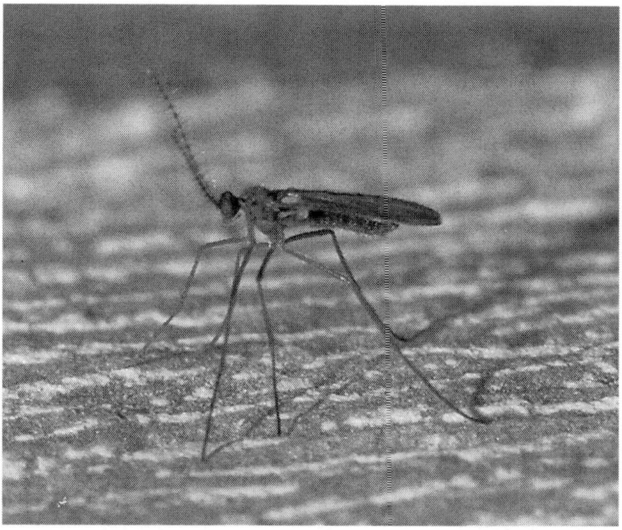

Figure 1.41: The gall midge, scientific name Cecidomyiidae.

Source: https://en.wikipedia.org/wiki/Cecidomyiidae

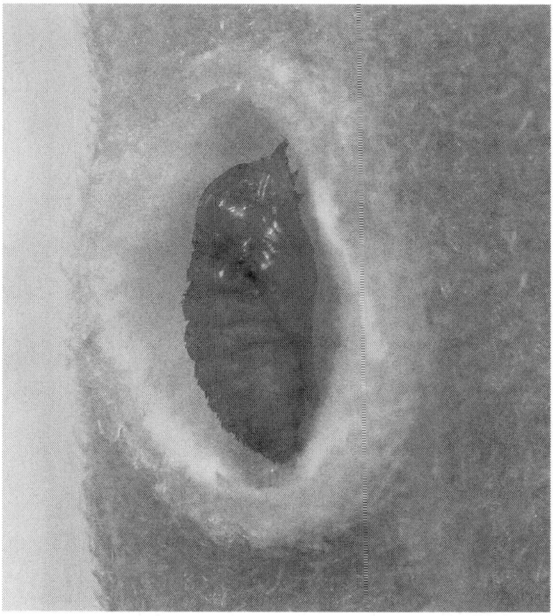

Figure 1.42: *Asphondylia solidaginis*, pupa in gall in Solidago species.

Source: https://en.wikipedia.org/wiki/Cecidomyiidae

Wasps which produce galls do this on their larval form especially on oak trees. Wasps which are tiny are the most common type of gall- forming wasps. They affect different regions of the plant such as the leaves, stems and twigs. The plant galls mostly develop after the female lays its eggs and the larvae produced are feed on the galls formed. The galls also offer protection to the larvae that are developing from parasites but some organisms sometimes penetrate this defense mechanism. They also occur on roses. Some of the gall types include the jumping oak gall, the oak apple gall, the oak bullet gall and stem galls found on the rose (Aloian, 2018).

Figure 1.43: The gall wasp.

Source: https://en.wikipedia.org/wiki/Gall_wasp

Figure 1.44: *Diplolepis rosae*, rose bedeguar gall.

Source: https://en.wikipedia.org/wiki/Gall_wasp

Figure 1.45: The oak artichoke gall.

Source: https://en.wikipedia.org/wiki/Gall_wasp

Despite there being a variety of species that produce gals, there are several limited species that can be attacked by gall- forming insects. The types of galls can also be categorized into different classes. They include the blister gall, bud galls; bullet galls produced by gall wasps; flower galls; fruit galls on both fruits and seeds; leaf galls; poach galls; root galls and many more.

Gall producers can infest an entire plant especially those that are young and still growing. Mature plant tissues are not affected by organisms that induce the formation of galls. The growth of the gall continues to grow with continuous feeding of the insect. If they start forming, they continue to form even if the insect dies. Insects feeding on the plants produce the galls. Feeding by the insect stimulates the growth of the plant through increased hormonal production and these create a form of deception which causes the plant to grow in a way that is not normal leading to the formation of the galls. Variations in the mechanisms used to form galls will vary depending on species of the gall producing insects (Aloian, 2018).

Some of the gall producing insects will infest and attack different parts of the plant host without affecting other plant structures but they can sometimes cause deformities which comes from the failure of certain plant organs being unable to have normal developments. The galls produced can be present in stems that are normal and can lead to little or no enlargement of the part of the plant which has been attacked. Galls that are indeterminate form deformities through stimulation or plant cell irritation where the gall producers feed. The cells that are close to where the insect is feeding are affected.

1.10 FLOWERS FOR A GREAT HOME TO SOME INSECTS

Flowers rely on insects for reproduction and the pollinators depend on flowers for food. Insects found in flowers play different ecological roles, some of them are pollinators and dispersal therefore their role is beneficial while others play the role of pests and parasites. There are a variety of insects found in flower ecologies.

Beneficial insects include bees and butterflies and they pollinate plants by the spread of pollen from flower to flower. They also spread microbes that benefit the ecology and surround the tissues of plants and they reduce the rotting of fruits for example strawberries these is attributed to Trichoderma fungi.

Figure 1.46: A bee extracting nectar from a flower.

Source: https://outerspacesgd.wordpress.com/tag/garden-habitats-for-insects/

The flower bug is also known as the minute pirate bug with about four hundred species. They lay their eggs in the tissues of plants and then the adult spend winter in plant debris. Most of the species feed on aphids found on flowers.

A variety of aphids feed on roses and the species that have commonly been identified is the rose aphid. They have a small body size. They have bodies that are soft, a body that has a pear shape; their bodies can either be green or pink and they are found in groups on different parts of the rose plant. Aphids consume sap. If the population is small less damage is caused to the plant. Their reproduction is rapid and large attacks reduce the quality of blooms in relation to flowers. They release honeydew as they feed attracting ants and wasps. These results to aphids and ants establishing relationships with are mutual aphids provide ants with honeydew and then the ants, in turn, protect aphids (Aloian, 2018).

Figure 1.47: *Macrosiphum rosae* the rose aphids on a rosebud.

Source: https://en.wikipedia.org/wiki/Macrosiphum_rosae

The leafhoppers are green insects that inject toxic substances into the foliage of flowers. This causes damage to the entire plant leading to damage of leaf tips and edges. They also spread the aster yellow virus. Whiteflies feed on plants and they secrete honeydew creating mold that is sooty and turns the plant yellow. They carry plant viruses and they are mostly found in greenhouses. Scale insects are garden pests that feed on flower plants through their entire growth period. They lead to growth that is stunted, dropping leaves and yellow spots on leaves. Thrips also feed on roses two of the most common thrips are flower thrips and western flower thrips. They suck the sap from plants and therefore feed on both flowers and leaves damaging them. Other insects that feed on rose plants include the leaf -cutting bee, grasshoppers and the Japanese beetles (Aloian, 2018).

Figure 1.48: Thrips on roses.

Source:https://www.greenlife.co.ke/thrips-on-roses/

1.11 KITCHEN PANTRY AS HOME TO SOME INSECTS

Food in our homes is sometimes stored in rooms called pantries. Some of the foods stored there include flour, cereals and other food products that are dried. Insects can infest or live in such rooms and feed on these stored foods. Different kinds of insects can be found in pantries. Some of the insects are found in pots, pans and window sills. The insects come from the food stores while others will come from the environment. They lead to food contamination. They can be controlled in a variety of ways for example using good containers, using insecticides, keep storage areas clean and organized and many more.

All insect stages may be present in the infested food packages or products. Common pests or insects found in pantries include moths, weevils, beetles, ants and cockroaches. Indian meal moths *Plodia interpunctella* commonly infest pantries in homes. They are sometimes found in products they have infested and a silky webbing is present on the product surface. They infest stored food that is dry such as cereals. Only the larvae feed in the stored product, upon maturing they leave the product and move for a distance before building a cocoon on walls or ceilings.

Figure 1.49: The Indian meal moth causing damage to sunflower seeds.

Source: https://en.wikipedia.org/wiki/Indianmeal_moth

Meal moths, *Pyralis farinalis* are less than one inch. They have front wings with a band that is red-brown across the top and bottom of the wings and also has a band with an olive- like color. The larva has a white body with a black head and an orange color at the end of the body. They feed on different flour varieties and grain products. They are not common in homes.

The saw-toothed grain beetles *Oryzaephilus surinamensis* have a red-brown color. They are found in a variety of fruits, cereals, nuts, dried meat, macaroni and seeds. Drugstore beetles feed on plant products that have been dried such as spices, macaroni and other foods that are grain- based. They also feed on tobacco products, paper products and dried flowers.

Figure 1.50: The saw- toothed grain beetle.

Source: https://en.wikipedia.org/wiki/Oryzaephilus_surinamensis

Figure 1.51: The saw- toothed beetles feeding on corn.

Source: https://www.controlling-pests.com/grain-beetle.html

Flour beetles,*Tribolium confusum* is commonly found in pantries. They are found in food products that have been dried such as bran, flour, cereal products, dried fruits, and chocolate. Warehouse and cabinet beetles feed on grain products, seeds, animal byproducts and pet food. Spider beetles infest plant products that have been dried. There are a variety of weevils that infest pantries such as the granary and maize weevils, rice weevils and been weevils.

Figure 1.52: The flour beetle.

Source: https://en.wikipedia.org/wiki/Flour_beetle

1.12 CLOSETS AND BOXES FOR A GOOD HOME TO SOME INSECTS

Fabric pests such as carpet beetles and cloth moths look for dark places where they can lay their eggs and this is also determined by the existence of food resources which the hatching larvae can feed on. The larvae cause damage to clothes, rugs or upholstery. They consume on any animal -based material such as wool, feathers, fur, silk and also lint. They do not consume materials that are synthetic but on mixes.

The carpet beetle has larvae which have a preference for dark places and are similarly secluded. They can also be found beneath furniture and dressers. They sometimes live without being noticed by people but live skin castings that are brown in closets, storage boxes or on clothes placed in closets. They shed their skin as they mature to adults which have a dark appearance with spots. Their larvae are cream in color. They can be controlled by cleaning, throwing out infested cloth material and constant vacuuming (Kalman, 2009).

Figure 1.53: Carpet beetles.

Source: manual.museum.wa.gov.au/conservation-and-care-collections-2017/appendices/appendix-4-insect-pests

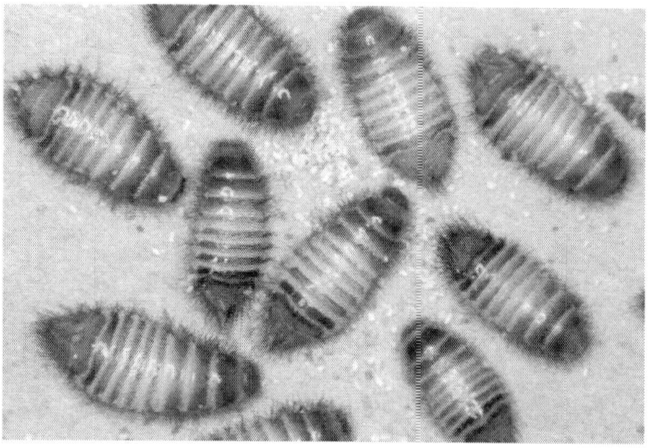

Figure 1.54: Larvae of the carpet beetle.

Source: manual.museum.wa.gov.au/conservation-and-care-collections-2017/appendices/appendix-4-insect-pests

For cloth moths, the larvae feed on fabrics which are woolen and silk. They remain unnoticed until there is widespread damage. They have a cream-white color. They seek dark areas for shelter. The adults do not use clothes for consumption but they lay eggs on the fabric which hatch into larvae. They are responsible for the creation of holes in clothing as they feed on the clothes. When the population infesting is large, they stay under upholstery or in heating vents (Kalman, 2009).

Figure 1.55: Cocoons left by the cloth moth.

Source: manual.museum.wa.gov.au/conservation-and-care-collections-2017/appendices/appendix-4-insect-pests

2

Aquatic Insects

CONTENTS

2.1 INTRODUCTION

These are insects that live in water. Water is their area of residence. For some insects, they get to spend part of their life in water more so while they are developing. For the aquatic insects, they are just like other insects, the only difference being that unlike the rest that lives on land, they live in water. Part of the aquatic insects is those that have the ability to dive in the water allowing them to feed off underwater things. Most of these insects tend to live in freshwaters where the oxygen supply is relatively higher compared to salty water which is known to have less of the oxygen supply. For these insects, their ability to thrive in this kind of environment, they are required to have special features that will allow them to breathe and move among other things.

The environment tends to use its components to allow it to thrive. This is still in the case of the aquatic insects. They are needed by the water in the processing of organic matter, the cycling of nutrients as well as the prey for aquatic -terrestrial and also aerial predators. This creates a balance in the ecosystem. Most of the insects have their life stages in the water this may include mayflies, stones, dragonflies, megalopterans and caddisflies among others. Some insect species are purely aquatic such as the beetles, bugs and the neuropterans among others. These insects are all specially adapted to these environmental conditions.

2.2 FEATURES OF AQUATIC INSECTS

For these insects to survive in this type of ecosystem, their bodies have to be specialized to suit this environment. They have most of the body parts common to other insects such as the legs, exoskeleton, eyes and antennae among others. For the adult aquatic insects, they have functional wings which may not be present in the juvenile stages of growth. The adult aquatic insects different from other insects as they tend to be long and slender, a good example is the dragonflies. Their light weight allows them to easily whiz around the water. For some of them, they have a fairly flat body allowing them to have an easy time moving in the water. These insects tend to have their breathing modified in a certain manner (Del-Claro & Guillermo, 2019).

2.2.1 Insect Breathing

Breathing involves the intake of oxygen and the release of carbon dioxide. For the aquatic insects, they need to have a modified breathing system that allows them to properly breathe underwater as they may not necessarily have direct access to oxygen from the atmosphere. Oxygen is an important aspect of living as it allows a variety of processes to be done in the body such as digestion of food. It is also important for insects to survive. Usually, for insects, their breathing system is usually simple. They usually draw air into the body through the spiracles. Spiracles are holes located on the sides of the abdomen. The spiracles are usually connected to the tracheal tubes allowing the absorption of oxygen. The oxygen enters the body of the insect through diffusion (Del-Claro & Guillermo, 2019).

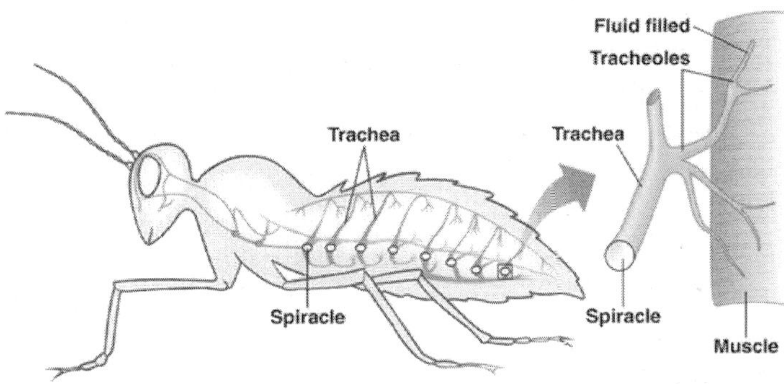

Figure 2.1: Insect's breathing system.

Source:https://quizlet.com/181009412/insect-internal-anatomy-respiratory-system-flash-cards/

In the case of the aquatic insects, there are some modifications made by the insects allowing them to live in the given environment. Some of the modifications should facilitate a faster diffusion so that they can breathe better (Heckman, 2008).

2.3 AQUATIC MODIFICATION

Some of the modifications include simple diffusion that is conducted over a relatively thin integument. This allows the air to enter into thebody at a faster pace. For some of them, it has involved the use of an air bubble. There is also the extraction of oxygen from the water through the use of a physical gill also known as a plastron. Some of the modification also includes the storage of oxygen in the hemoglobin molecules in the hemolymph. There is also the taking of oxygen from the surface through breathing tubes also known as siphons.

Various modifications are required and needed in the breathing conducted by these insects. They are explained as shown below:

2.3.1 Circular Respiration

To facilitate breathing, most of these aquatic insects have relatively thin integuments which are permeable to oxygen molecules as well as carbon dioxide. These gases are involved in the process of breathing. In breathing, it involves oxygen diffusing into the body of the insect while carbon dioxide diffuses out the body. The diffusion of these gases through the thin integuments also known as body wall may be sufficient to meet the needs of the metabolic demands of the small inactive insects more so those that live in the cold fast- moving streams that have plenty of oxygen. For adult insects, they are able to move to areas with less oxygen supply and make use of the physical gills to allow them to survive in these conditions (Heckman, 2008).

2.3.2 Biological Gills

This is another modification of the insects. A biological gill is an organ present in most of the aquatic animals that facilitates the dissolving oxygen from water to pass into the body of the organism. In insects, these gills are some of the modifications that allow them to survive in water. These gills are usually outgrowths of the tracheal system. These gills are covered by a thin layer of cuticle. This cuticle is also permeable to oxygen and carbon dioxide. Some of the insects that have this kind of gills include mayflies and damselflies. The gills are usually in a leaf-like manner. It is also located on the sides or rear of the abdomen. The fanning movements of the gills facilitate them being in contact and there is a constant supply of freshwater.

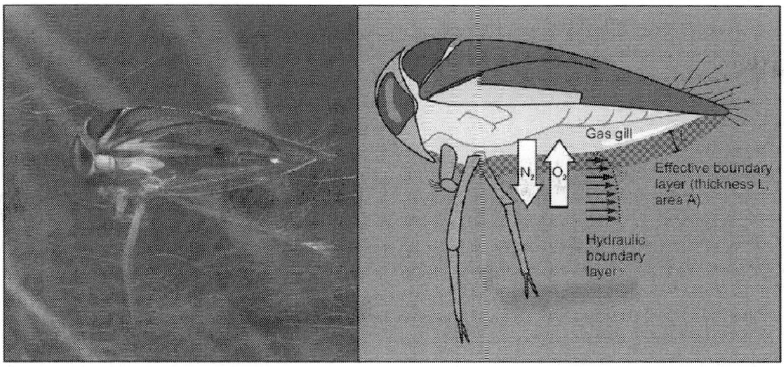

Figure 2.2: Showing biological gills in insects.

Source:https://jeb.biologists.org/content/216/2/164

In the case of stoneflies and caddisflies, they have the filamentous gills located in the thorax or the abdomen. The dragonflies are different from aquatic insects in that they have internal gills that are associated with the rectum. The water is circulated in and out of the anus. This is facilitated by the muscular contractions in the abdomen. This mechanism is known as the rectal gill mechanism. This acts as a double jet propulsion system. The contractions by the abdomen result in the jet of water and it leads to the insect being thrust forward allowing the insect to move forward and escape predators (Jong, 2018).

2.3.3 Breathing Tubes

These are also among the modifications of the aquatic insects. Some of the aquatic insects have the ability to get air directly from the atmosphere without directly obtaining it from the water. This is achieved through hollow breathing tubes that are also called siphons. They also follow the principle of the Driver snorkel. In the case of the mosquito larvae, they have a siphon tube which is an extension of the posterior spiracles. At the opening, there are rings of closely spaced hairs that have waterproofing coating that guard the siphon. In the air-water interface, in this area, the hairs break the surface tension of the water. This also ensures that the airway is open. When the insect dives in the water, the water pressure pushes the hair close together. This facilitates the sealing of the airways keeping the water out.

Figure 2.3: Showing breathing tubes in insects.

Source:https://ibycter.wordpress.com/tag/aquatic-insects/

Other examples of insects that have snorkel-like breathing tubes include water scorpions and rat-tailed maggots. Some insects insert their breathing tubes into their air stores. By so doing they are able to obtain a rich supply of oxygen without them having to swim to the surface of the water.

2.3.4 Air Bubbles

There are aquatic insects that have the ability to carry a bubble of air as they dive into the water surface. Some of them include diving beetles. This bubble may be held underwing covers also known as elytra. In other cases, it may be trapped against the body through the use of specialized hairs. The bubble usually covers one or more spiracles of the insect. This allows the insect to breathe even when it is submerged in water. The air bubble may supply the insect with oxygen for only a short period of time. In more specialized breathing systems, it allows for a short term oxygen supply of oxygen while in the water.

Insects have unique physical properties allowing the bubble is able to collect more oxygen molecules that have been dissolved in the surrounding water. In this manner, the bubble acts like a physical gill. This is because it replenishes the supply of oxygen through passive diffusion. The larger the surface area of the bubble the higher the chances of the system working efficiently. This allows the insect to remain underwater for as long as possible. This is in the condition that the volume of the oxygen that is diffusing into the bubble is much greater than the amount of oxygen that is being consumed by the insect. The downside is that gradually the size of the bubble will reduce as the nitrogen diffuses into the water (Rufer, 2015).

It is important to note that the bubbles surface area decreases then the rate of gas exchange will also decrease. This will lead to the bubble becoming too small to meet the metabolic demands. This means that the insect will need to replenish the water supply by going to the surface. They are then able to get a new supply of oxygen into the system (Rufer, 2015).

2.3.5 Plastron

This is another mechanism present in aquatic insects. The plastron is a unique array of rigid that is closely-spaced hydrophobic hairs that facilitates the creation of airspace located next to the body. The air that is trapped within a plastron usually operates asa physical gill. This is a special case as the airspace cannot shrink in volume. This is attributed to the fact that the fortress of hydrophobic hairs that prevent the encroachment of surrounding water. As the insect consumes oxygen it leads to the creation of a partial pressure deficit that is within the plastron.

Figure 2.4: A plastron in insects.

Source: https://www.mentalfloss.com/article/78685/how-do-water-beetles-breathe-underwater

This deficit may be corrected by dissolving oxygen that diffuses out of the bubble. Also as nitrogen gradually diffuses from the bubble, it leads to the creation of a partial pressure deficit. In the case where there is little dissolved nitrogen in the water some of the partial pressure caused by nitrogen, the pressure deficit is corrected by oxygen. In this manner, the plastron makes a trade on the nitrogen in exchange for oxygen by keeping a constant volume of gas which may gradually become enriched with oxygen. The constant volume of the air supply in plastron removes the periodic need to surface and replenish the bubble. The insect will ,therefore, surface for some time. The sizes of the bubble will a major determinant in knowing how many times the plastron will visit the surface (Nair, Morse & Marshall, 2015).

For insects that tend to be permanently submerged such as the riffle beetles and the family, Elmidae tends to have plastrons. The same case also applies tothe insects that cannot reach the surface. Good examples are the egg of floodwater mosquitoes. While underwater the structures are seen as thin, silvery films of air covering parts of the body surface. This mechanism has been useful in most of the insects.

2.3.6 Hemoglobin

This is a component of blood that is present in humans and also in insects. It is a major component of blood that plays amajor role inthe transportation of oxygen. This pigment facilitates the capturing of oxygen molecules allowing it to enter the body. This molecule is also used in the carrying of carbon dioxide and when this component is close to the breathing surface where the carbon dioxide molecules are removed. This is an essential component of blood thatis not commonly present in most insects. These molecules are usually present in larvae of certain midges which are known as bloodworms. Thesetypes of larvae tend to live in muddy depths of ponds or streams where the dissolved oxygen may be of feweramounts than what is required.

In normal conditions, the hemoglobin molecules in blood bind and may have a reserved supply of oxygen. In the case where the conditions become anaerobic, the oxygen is gradually released by the hemoglobin. The released oxygen is utilized by the cell and tissues of the body. This is a form of a back-up supply that may last for a few minutes. The advantage is that it may provide oxygen for a period that is enough for the insect to relocate to a more oxygenated area.

2.4 TYPES OF GILLS IN AQUATIC INSECTS

2.4.1 Tracheal Gills

This involves insect breathing through the tracheal gills which are a network on tubules. The relatively large tubules are usually connected to the spiracles that are opening at the surface of the body facilitating the entry of oxygen into the body and the removal of carbon dioxide from the body. As stated above the spiracles are located on the pleural surface of the body. They are usually located on each side of the individual segments. As the airflow may be tidal, some of the airflow is directed into the anterior spiracles and isreleased through the posterior abdominal. These are interconnected with the longitudinal and transverse tracheal trunks which make the flow to be directed making it more efficient. Compared to the use of the tidal flow this system is more efficient as it is constantly flushed and the entering air is not mixed with the air that is used.

The larger tracheal tubes usually send of branches that end up becoming smaller in diameter as they ramify to all tissues. In most cases, the smallest size of the diameter of the tracheoles ends blindly on the surface of the cells. In some cases, it indents the cells until they terminate. There are subdivisions in the tracheal gills which include the primary tracheation of the tracheal gills. There is also the paraspiracular tracheal gill which is usually paired to the abdominal appendage. For these gills the trachea arises may be from the spiracular trachea of the same abdominal segment which is from the nearest point of the tracheae branching. For these kinds of tracheal gills occur independently in aquatic larvae of some insects. As evolution has occurred the paraspiracular tracheal gill has occurred as a respiratory organ. It substitutes the primary spiracle as well as the tracheated which is from the primary spiracular trachea (Del-Claro & Guillermo, 2019).

The tracheal gills are usually analogous and are not homologous organs. This is because they occur independently in non-related taxa and perform the same function. When discussing the tracheal gills, it is important for one to discuss the tergalius as most people tend to confuse them. These are paired abdominal appendages of a peculiar structure. It commonly occurs in the larvae of ephemeroptera. They usually have a commonplace of attachment which is on the posterior-lateral corner of the abdominal tergite and is constantly moved by the special tergalial muscles which are constantly in the abdomen. When compared to the trachea gills, the tracheation of the tergalii is usually secondary. Instead of the tergalial arising from the spiracular trachea, it arises from the longitudinal tracheal trunk. The tergalii are usually homologous but not analogous organs. This is attributed to the fact that they are of the same origin but perform the same function (Del-Claro & Guillermo, 2019).

2.4.2 Spiracular Gills

This is a pupal structure. Its main duty is to be the main respiratory organ in the case of species of the adult taphrophila. The structure is present before the cuticle is shed. In the phase where the pupa-adult is molting, the epidermis as well as the blood in the spiracular gill are usually completely isolated from the living insect. This is done by the two cuticles present in between the molting fluid. After the isolation of the tissue, the epidermis of most parts of the gills starts to dissociate. After this occurs, the cells become rounded and detach from the cuticle and from one another. In the process, they form loose clumps that are far from the gill walls. The tissues found in the gills repairs, cuts and tears in the gill walls with a tanned cuticle.

The competence of the isolated tissue to repair the damage done to the gill walls may last about 14 days. The tissue is then isolated for 8 to 9 days before the adult emerges. There is the repair of the gills for 5 days after the insect has shed it and flown away. The isolated tissues of the gills are able to tolerate dehydration and high temperatures. The epidermis also has the same characteristics. This structure is not effective. In most cases, the spiracular gills are either a modification of the spiracles or of the body that is adjoining the spiracle or in some cases, it takes both the wall and the spiracles. The spiracular gills occur mostly in the larvae stage. The spiracular gills are usually present in insects that have a plastron or an opening that allows oxygen into the body and it diffuses through air channels.

2.4.3 Physical Gills

For most insects, the presence of the physical gills is the thing that allows them to submerge in water preventing them from drowning. There is a pressing need for thesegills because, for these insects, they tend to have a slow diffusion of oxygen into their system which cannot allow them to meet the respiratory demand by the body for various processes. Therefore, most of the insects have physical gill which allows them to breathe properly while underwater. This has made this type of breathing to be known as the plastron respiration. This is because it demonstrates similar characteristics to the respiratory system of insects that make use of the plastron (Heckman,

2008).

This gas film allows an easy interface for oxygen to enter into the body of the insect. This system is modified in such a manner that an efficient amount of oxygen gets to enter into the body of the insect that is enough for the body to deal with. They provide a large gas-water interface to facilitate the entry of enhanced uptake of oxygen into the bloodstream while underwater. This is in the condition that the spiracles provide that gas-phase in contact with the water. The use of these gills allows the survival of insects while underwater.

2.5 FEEDING

A major part of living is the ability to acquire and feed on food. Every creature needs food for it to survive. Food is an essential part of the living process as it is the source of energy and nutrients needed by the insects allowing various body processes to be conducted. This means that the mouth as well as the digestive system have to be modified for it's to suit the given environment. The diet of these animals hasbeen modified so as to adapt to the given environment. The feeding of these insects usually occurs at the nexus of the habitat. The food source, size and behavior aredependent on the complexities of the mouthpart morphology. The feeding habits may also vary depending on the stage of development. This implies that there are a variety of feeding groups (Rufer, 2015).

Being that these insects live in water, their probability of them being eaten by the fish is very high. For these insects, they need to be extensively modified to escape from danger or else they will be prey to most of the aquatic life. In the aquatic environment there a variety of predators. The mouthparts of these insects are modified to feed in these environments. There a variety of foods that are eaten by insects. For the immature insects, they feed using a variety of ways that may involve the feeding on suspended particles known as seston that are present in the water column. The mode of the representative of a variety of insects.

Depending on the species of insects we look into the food sources, particle sizes as well as the feeding behavior of the insects. The feeding of these insects is important as it affects various factors of the environment such as the primary production achieved through grazing, there is the detritus breakdown, nutrient mineralization and downstream spiraling. There are functional feeding groups that are present that are developed as a tool that allows for the incorporation of macroinvertebrates when studying the aquatic ecosystems. The classification is done and having the advantage of the morphological characteristics such as the mouthpart specialization and the behavioral mechanisms (the way that they feed these food particles).

When they get to eat these foods they lead to the breakdown of most of the nutrients that are available in the environment. This contributes to the breakdown of nutrients in the environment. There is a need for us to get an understanding of the functional feeding groups (Rufer, 2015).

2.5.1 Functional Feeding Groups

These groups were developed by Cummins and have been developed by several ecological projects. The groups have been used in the development of various concepts while in different temperate and tropical conditions. The groups were made to properly understand the role of the macroinvertebrates play in the functions present in the ecosystems which include primary production, detritus breakdown and downstream spiraling. Some of the groups are shown below:

Scrapers: These are organisms that consume resources that grow over substrates by removing them. This is by removing them with their parts that have been adapted to crop the particle that have been closely attached to the rocks as well substrates. These species consume a variety of resources such as algae attached to rocks. They also feed on benthic biofilms that entail bacteria, algae, fungi and polysaccharides. The polysaccharides usually cover the hard substances in the aquatic ecosystems. These kinds of feeders are usually present in the stream ecosystem. The fact that they consume algae means that there will be an impact that will be felt by the producers. They consume the producer biomass and they also affect the algal species compositions. It is important to note that the variety in the densities of the scrapers that have led to a variety of effects that range from positive to negative on the algal biomass as well as its production.

Piercers: These are organisms that feed on vascular plants. This is done through cutting or piercing the tissue by using either the sharp or chewing mouthparts as well as ingesting the plant liquids. The small caddisfly species that are known to pierce individual algal cells in consuming cellular fluids hence avoiding the consumption of the cellulose-rich cell walls.Thisgroup was not majorly considered till later on after a few studies were conducted on the piercing macroinvertebrates. They may include organisms such as the larval Hydroptilidae that tend to be commonly present in streams and are anticipated to have important effects on the algal community.

Shredders: There are organisms that either cut or chew pieces of living or dead plant material such as the plant parts. The major role of the shredders is to aid in the breakdown of the large particles of plant materials into relatively smaller pieces thatare to be transported downstream. The shredders are organisms that avail the nutrients to the microbial consumer. They are in most cases referred to as the consumers in the coarse particulate organic matter and result in the production of the fine particulate organic matter. The organisms that feed on plant tissue are responsible for the losses in the aquatic vascular plants. The insects known for these kinds of feeding are the chrysomelid beetles which focus more on the vascular plants. Some shredders consume detritus and have a relevant effect in the environment as they are said to increase the amount of fine particles in the streams.

Collectors: Under this category, there is the collector and the collector-gatherers who are organisms that have been modified such they have mouthparts that facilitate the sieving and collection of small particles that have formed on the stream bottoms. They bear a resemblance to the shredders in that they have the ability to consume small pieces of leaves. However, their mouthparts do not have the ability to cut the leaves into smaller pieces. They are therefore able to consume those that are small in size. These groups of insects plays a major role in re-packing the fine particulate organic materials into larger particles after they have ingested them.

2.5.2 Insect Mouthparts

When it comes to feeding it is important for one to know the mouthparts of the insects so as to get a better understanding of their feeding behaviors. For the insects living in water, their mouthparts have been adapted to suit the given environment. These insects have mouthparts to facilitate cutting, chewing, piercing, sucking, shredding, filtering and siphoning. There are a variety of insect mouthparts. Through the years insects have undergone evolution so as to facilitate their feeding habits. Studies have shown that these modifications that have been occurring throughthe years. A great illustration is the with the mosquitoes as well as the aphids whose mouthparts have been catered to allow for piercing and sucking but the modification has facilitated the female mosquitoes being able to feed on blood.

The following are some of the insect mouthparts. They include:

Labrum: This is a mouth part that is an extension of the head that covers the mandibles. Different from other mouthparts, theseparts usually occur as a single fused plate. Initially, it was and embryonically is two structures. This structure is located on the midline and is the upper-most mouth part. Its major function is to hold food in place as the chewing process occurs and conducted by the mandibles. In some cases, it is referred to as the upper lip.

Mandible: This is another part of the mouth of insects. Insects that chew are said to have two mandibles each at the sides of the head. It is said to be the largest part of the insect's mouth. This part is used in the mastication of food materials. Mastication refers to the cutting, shredding, crushing and chewing of food. During the process, the mandibles open up outward in the sides of the head and they come back together medially. In some cases, the mandibles are referred to as jaws and are presented in diverse ways depending on the type of insects. In the case of the beetles who are among the aquatic insects the mandibles vary depending on the feeding habits. Some beetles live in the water and those that live on land while others live on both. In the case of the carnivorous beetles, their beetles are usually extended to allow them to properly seize their prey. In the case of the tiger beetles, their mandibles have been specialized to allow them to feed on their prey.

In the case of the diving beetles, their mandibles have been modified in that they are hollow allowing them to ingest digestive fluids so that they are able to liquefy the tissues of their prey. They are then able to suck in the digested tissue by using their mandibles.

Maxilla: This mouth part is utilized in the cutting and manipulation of food during the process of mastication. These parts have hairs as well as teeth along the margins. The margins on the outer area arethe galea which is a cupped or scoop- like structure. It rests on the outer edge of the labium. This mouth part is specialized in that it has palps which allow the properties of potential foods.

Labium: This is usually a single structure that was formed through the fusion of two secondary maxillae. In most cases, this moth part is described as the floor of the mouth as well as working in close with the mouth of the insect. This mouth part aids in the mastication process. This part in some of the aquatic insects it is modified to perform different functions such as the case of the jaws of the nymphs of the Odonata among other insects. For them, the labium are folded neatly beneath the head. They have been modified in that the insect can flick it out to allow it to snatch prey lifting it to the back of the head. After this is done, the chewing mouthparts demolish the prey and afterward swallow it.

This mouth part is attached to the cibarium. Its broad basal portion is divided into submentum which is the proximal part of the mentum present in the middle. The other division is the prementum at the distal section. It contains a structure known as the ligula which entails the inner pairs of lobes referred to as glossae while the lateral pair is the paraglossae.

Hypopharynx: In the case of the labium we get to discuss the hypopharynx which is like a globular structure. This is because it is located at the base of the labium. It is found next to the mandibles and the maxillae. It is said to be related to the salivary glands and aids in the swallowing of food. In the siphoning insects, this mouth part is used in the sucking of fluids. This is to be done without the piercing of food, sponging or licking. There are variations when it comes to some cf the insects. There is also the proboscis which is present in the adult Lepidoptera. This is a long tube that was formed by a pair of the galeae of the maxillae. In the case of the large vertebrates, these parts perform roles done by the tongue.

These are the mouthparts present in insects and most of them are possessed by the aquatic insects. This allows them to feed depending onthe type of food that they ingest. When discussing the feeding of insects, we get to look into its digestive system. This is basically the digestive system of insects.

2.5.3 Insect's Digestive System

Like human beings, insects have a digestive system that allows for the extraction of nutrients.

For the case of insects, the food is ingested in the form of macromolecules as well as other complex substances that need to be broken down through catabolic reactions. The complex substances include fats, polysaccharides, nucleic acid and proteins. After these substances have been broken down, they are used by the body for the provision of energy, facilitating growth or being utilized in the process of reproduction. The above- described process is known as digestion.

In the case of insect digestive system, it is closed. There is one enclosed coiling tube referred to as an alimentary canal. This system allows for the entry of food in the mouth as it progresses inside the body until it leaves through the anus. In the alimentary canal, there are special regions that have been specialized for given functions such as grinding, storage of food, nutrient absorption and enzyme production. There is also the presence of sphincters that facilitate controlling the movements of foods and fluids within three major regions. They include stomatodeum, mesenteron and proctodeum (Jong, 2018).

Insects however have paired salivary glands and reservoirs. These structures are located in the thorax region. The salivary glands are used in the production of saliva. The ducts lead from glands to the reservoirs. They then head to an opening referred to as salivarium located behind the hypopharynx. The movement in the mouthparts aid in the mixing of food while in the buccal activity. The saliva then mixes with food. The resultant travels into the mouth through the salivary tubes. This is the initiation of the digestion process. Two regions including the stomatodeum and the proctodeum are invaginations of the epidermis. They are lined with intima. In the case of the mesenteron, it is not lined with a cuticle but instead with it has rapidly dividedand are constantly replaced with epithelial cells.

With the consequent molting along with the exoskeleton, the cuticles are shed. The food moves down the alimentary canal through a process known as peristalsis. Below is a vivid description of the regions of the alimentary canal (Jong, 2018). They include:

Stomatodeum; It is also known as the foregut. This is the region involved with the storage, grinding and transportation of food to consequent regions. Some of the parts in this region include the buccal cavity, the pharynx, the esophagus, the crop and gizzard. There is also the salivary secretions from labial glands that dilute the ingested food. In the case of mosquitoes who are insect blood- feeding insects, the blood thinners are anticoagulants are released in this region.

Mesenteron: It is also known as the midgut. The digestive enzymes in this region are generated and secreted into the lumen. It is followed by the nutrients being absorbed into the insect's body. In this region, the food is enveloped as it arrives from the stomatodeum by the peritrophic membrane. This is a mucopolysaccharide layer secreted from the midgut's epithelial cells. It is said that this membrane prevents food pathogens from contacting the epithelium as well as attacking the insect's body. It functions as a filter allowing the entry of small molecules and hindering large particles from entering and reaching midgut cells. The large substances are broken down into smaller ones, digestion and absorption of nutrients occur at the surface of the epithelium. There are microscopic projections known as microvilli thereby increasing the surface area for absorption in the region. This allows maximum absorption of nutrients.

Proctodeum: Also known as the hindgut. This region is divided into three parts which arethe hilum, colon and rectum. The interior section is the hilum while the needle section is the colon. This extends from the pyloric valve that is found between the mid and the hindgut to the anus. Processes such as absorption of water, salts and other substances is done before excision. Insects are similar to other animals in that the removal of toxic metabolic waste requires water. However, in insects, water conservation is important contributorto the need forblind -ended ducts known as Malpighian tubule. The ducts immerge as invagination's located at the anterior and of the hindgut they are major organs of excretions and osmoregulation. They obtain waste products from the hemolymph. The tubules produce the insect's uric acids which are then transported to the hindgut allowing the re-absorption of water and salts by the hindgut and rectum. The excrement is avoided as non-toxic and insoluble uric acid granules. Osmoregulation and excursion in insects arenot done by the mulligan tubules alone as they require a joined function of the rectum or ileum.

The digestive system of the insects maybe modified according to the type of food they feed on.

2.6 LOCOMOTION IN AQUATIC INSECTS

When we talk about insects in a water environment we need to understand how their body is adapted for this kind of environment. Insects tend to have fixed exoskeletons, which imply that their bodies are rigid and firm. Insects are known to make use of their legs and wings. These body parts have to be modified to meet the needs of the ecological environment. They need to be buoyant therefore allowing them to float on water. For some of the insects, they have water jets that allow them to make the needed movement while in the water. This allows them to propagate from one area to another. For others, they make use of the undulatory movements of their soft body parts (Heckman, 2008).

The most commonly used mechanism in locomotion is the use of application of pressure against water. Insects like other animals are able to attain locomotion through the use of contact with the substrate. The locomotion in insects is made possible through the thorax and the appendages. For the aquatic insects, they have jointed limbs thathave evolved in the marine. For us to know how insects' locomote we need to have an understanding ofthe insect legs.

2.6.1 Insect Legs

The legs of an insect are made of five parts which include the coxa, trochanter, tibia, femur and tarsus. Form the furthest part is the tarsus and it ends with a claw or a pair of claws. For some insects, there is a specialization in their legs in that they have special structures including seta, spines and dactyls. Generally, the insect legs are made up of three parts which include the forelegs, middle legs and hind legs. In the case of aquatic insects, their legs may be used in swimming and skating. This means that they use their middle or hind legs as oars allowing them to swim and dive. To facilitate this, their legs have been modified in that they are flattened. In some of them, the legs are equipped with a fringe of long stiff hairs improving their performance and efficiency while in water. In the case of larvae who are legless and pupae in mosquitoes, they are able to swim in the water by twisting, undulating or contorting their bodies.

For those that utilize jet projection for movement, they propel themselves by contracting their abdominal muscles and forcing a jet of water out of the rectal chamber that holds the respiratory gills. For insects that skate, their legs have been modified in that they have hairs at the tips of their feet preventing the leg of the insect from breaking the surface tension of water supporting their bodies (Heckman, 2008).

Insects tend to have three pairs of legs each of which is present in the given segment. The type of leg they have is dependent onthe type of locomotion as well as the type of prey they have. As the insect develops then the legs keep on being modified. As the insect develops from the larval stage to the adult stage, in the subsequent stages their bodies are modified to allow them to locomote in the water (Jong, 2018).

2.7 EXCRETION AND OSMOREGULATION IN AQUATIC INSECTS

Both excretion and osmoregulation are common phenomenain living beings both animals and insects. In the case of insects, their feces can either be in solid or liquid form and may be composed of undigested food and metabolic excretions. As we are discussing on aquatic insects, it is good to know that the type of waste may be affected by the nature of the environment which contains mostly water. In their case, their wastes are made of dilute wastes which arereleased to water by flushing with water. This means that their mode of excretion may be different from the terrestrial insects whose bodies may be adapted to conserve water. Aquatic insects need to conserve ions which include sodium, chloride and potassium. This may result in the limiting of the food which may be lost to water through the process of diffusion (Jong, 2018).

In understanding the process of the removal of urea in aquatic insects, we discuss more on excretion and osmoregulation which are the processes that are actively involved in the creation of wastes. They contribute to the maintenance of the conducive osmotic pressure and ionic concentration of the body fluids. The excretory system is the system in charge of excretion and osmoregulation. The activities done by this are conducted mostly by the Malpighian tubules and hindgut. In the case of the freshwater insects, the hemolymph composition is controlled responding to the loss of ions to the environment. The processes are done with the help of the excretory system and special cells. The special cells can also be called chloride cells. They are present in the hindgut. They have the ability to absorb inorganic ions present in the dilute solutions (Del-Claro & Guillermo, 2019).

For us to get a better understanding, we can look into the Malpighian tubules.

2.7.1 Malpighian Tubules

This is an important organ in excretion and osmoregulation in insects. Thesetubules work assisted by the ileum and rectum. They are outgrowths of the alimentary canal and entail long thin tubules that have been formed by a single layer of cells that surround a blind-ending lumen. The individual tubules are externally covered by a peritoneal coat. It is also supplied with muscle fibers and also tracheoles. There are a variety of functions conducted by these tubules. The major role may be the fact that they are utilized in the removal of unwanted wastes. Other functions include them being used in spittle secretion in the case of the spittlebug, there is light production in the case of Bolitophila and silk production in the larval neuropteran.

In the physiology of the Malpighian tubules include it producing a filtrate that is osmotic but isionically lack a resemblance with the hemolymph. It selectively reabsorbs water along with other solutes while eliminating others. They produce an isosmotic filtrate that is rich in potassium and low in sodium having chlorine as the major anion. There is active transportation of ions more so for potassium into the tubule lumen thereby generating an osmotic pressure gradient allowing the passive flow of water. Most amino acids and sugars are passively filtered from the hemolymph through various junctions that are present between the tubule cells. It is also the place where the amino acids as well non-metabolized and toxic organic compounds are actively transported into the tubule lumen. Sugar is usually reabsorbed from the lumen and returned to the hemolymph. There are secretion activities by the Malpighian tubules results to a flow of primary urine from its lumen towards and into the gut (Del-Claro & Guillermo, 2019).

While in the rectum, the urine is modified by the removal of solutes and water leading to the maintenance of fluid and ionic homeostasis of the body. In the case of nitrogenous excretion in aquatic insects' ammonia is the main excretory product which is freely soluble in water and may need more amount of water for the waste product removal. This type of excretion is also referred to as ammonotelism. In the case of cryptonephry, it is the distal end of the Malpighian tubules which are held in contact with the rectal wall by the perinephric membrane. This is concerned with the efficient dehydration of faces before ionic regulation or elimination (Nair, Morse & Marshall, 2015).

When discussing aquatic insects, some live in freshwater while there are those that live in marine water. There are also river insects. The river insects are those that live in rivers. Rivers are important ecosystems and provide good homes for insects. Rivers are homes to insects, fishes and plant species among others. This ecosystem is well adapted in that they are able to provide food for all the creatures living in these environments. This ecosystem must be protected to ensure that the insects survive. For so long the insects in these insects have been affected by the fact that there has been dumping of wastes in the rivers.

This has led to the insects being extinct. Some of the aquatic insects have been extinct due to the dumping. It has led to the introduction of hazardous substances into the water killing the insects. It has also led to the reduction of oxygen in the rivers. This has been identified through research that has been conducted when scientists conduct research on water. This has led to the need for the clean up to be done in the rivers. This is because for the insects they spend most of their time underwater while in the larva stage. They emerge from the water while in the adult stage having wings where they can mate with other insects and reproduce (Nair, Morse & Marshall, 2015).

2.8 REPRODUCTION IN AQUATIC INSECTS

Insects are known to reproduce through oviparously through laying of eggs. The eggs are produced by the females in their ovaries. The males produce the sperms through the testis. They are transmitted to the females through mating. The semen is deposited into the external genitalia. The sperms are stored in the female's body in the spermathecae. When fertilization occurs the eggs travel along the oviducts. They are then fertilized by the sperm and are expelled from the body such as through the ovipositor.

2.8.1 Internal Reproductive Parts

Females: In the case of females in the case of reproduction they need to be able to make eggs. Their bodies are adapted to receive and store sperms. They are suited to manipulate the sperms from different males and then they lay eggs. Their bodies are made up of a pair of ovaries, accessory glands, spermathecae and ducts that facilitate the connection of these parts. The ovaries are in charge of the production of eggs while the accessory glands are used in the production of substances that aid in the package and laying of eggs. The spermathecae areused in the storage of sperms for varying durations of time. This is alongside the oviducts which facilitate the control of the utilization of sperms. The ducts as well as the spermathecae are lined with a cuticle.

The eggs are usually made up of a number of egg tubes known as ovarioles that vary in size and number depending on the insect species. The eggs that are made insects are dependent on the number of ovarioles and the rate at which the ovaries develop which is influenced by the design of the ovaries. In the meroistic ovaries, the eggs are to undergo division repeatedly and most of the daughter cells become helper cells for a single oocyte within the cluster. In the case of the panoistic ovaries, the individual egg to be produced by the stem germ cell develops into an oocyte there may be no helper cell from the germline. There is the production of eggs by the panoistic ovaries that tend to be slower compared to the meroistic ovaries (Rufer, 2015).

There are the accessory glands also known as the glandular parts of the oviducts produce a variety of substances for sperm maintenance, transport and fertilization. This is also utilized in the protection of eggs. They can also produce glue as well as the production of substances for coating eggs or tough covering for a given batch of eggs that are referred to as the oothecae. The spermathecae are sacs that are used in the storage of sperms allowing them to be stored between the time of mating and the time in which the egg is fertilized.

Males: The main element for reproduction is the testis which is suspended by the body cavity by the fat body and the tracheae. There are more primitive insects that have a single testis while in the case of the lepidopterans there are two maturing testes and are secondarily fused into a structure during the later stages to facilitate the later stages of larval development although the ducts leading from them to remain separate. Most of the male insects have a pair of tsetse inside of which there are sperm tubes enclosed within a membranous sac. The follicles are connected to the vas deferens by the vas efferents. There are two tubular vasa deferentia that connect to a median ejaculatory duct that leads to the outside.

There is a portion of the vas deferens that is often enlarged to form the seminal vesicle that stores the sperm before they are discharged into the female during the mating process. There is the seminal vesicle that hasglandular linings that secrete nutrients which facilitate the nourishment and the maintenance of the sperms. There is the ejaculatory duct which is derived from an invagination of the epidermal cells in the process of development and ends up having a cuticular lining. There is the terminal portion of the ejaculatory duct that may be sclerotized to form the intromittent organ and the aedeagus. The remaining part of the male reproductive system is from the embryonic mesoderm withthe exception of the germ cells. They descend from the primordial pole cells very early duringembryogenesis (Rufer, 2015).

2.8.2 External Reproductive System

The abdomen of an adult insect is made of 11-12 segments that arerelatively strongly sclerotized compared to the head and the thorax. The individual segments of the abdomen have been represented by the sclerotized tergum, sternum and in some cases the pleurae's. The terga are separated from each other and theyform the adjacent sterna or pleural by a membrane. The spiracles are usually found in the pleural area. There is a variation of the ground plan involves the fusion of the terga or the terga and sterna resulting in the formation of a continuous dorsal or conical tube. In some of the insects, they have a sclerite in the pleural area known as a laterotergite. In some instances, the laterosternites can be utilized as the name for the ventral sclerite.

While in the embryonic stage of most of the insects and the embryonic stage of the primitive insects there are 11 abdominal segments. In the new insects, there is atendency leaning towards the reduction in the number of segments. However, the primitive number 11 is maintained in the process of embryogenesis. There is a variation in the abdominal segment number is considerable. In the case where the Apterygota is taken under consideration to be indicative for the ground plan for the pterygotes then confusion. The adult Protura are known to have 12 segments while the collembolan has 6 (Rufer, 2015).

The orthopteran family Acrididae has11 segments while the fossil specimen of the Zoraptera they have 10-segmented abdomen. In most cases the first seven abdominal segments of the adults which are similar in structure and do not have appendages. Some Aapterygotes are serially homologous with the distal parts of the thoracic legs. In the mesally there are one or two pairs of the protrusible vesicles on some abdominal segment. The vesicles are derived from the coxal as well as the trochanteral endites of the ancestral abdominal appendages. The aquatic larvae and nymphs may possess gills laterally on some to most abdominal segments. The other parts of the abdominal segments are composed of the reproductive and anal parts.

The anal-genital part of the abdomen iscomposed of the 8th and 9th segment to the abdominal apex. The 8th and 9th segments carry the genitalia segment 10 is seen as a complete segment in most of the lower segment but they do not have appendages. The 11th segment is represented through the use of a dorsal epiproct as well as a pair of ventral paraprocts derived from the sternum. There are a pair of appendages that articulates laterally on the segment 11. They are typically annuated and filamentous. However, they have been modified takes the case of the forceps of the earwigs. Some reductions have been noted in different insect orders. The annulated caudal filament arises from the tip of the Epiproct in the Apterygotes in most of the mayflies. There is a similar structure in the nymphal stoneflies which is of uncertain homology. The terminal abdominal segments contain excretory and sensory functions in all the insects. In the case of the adults, there is and additional reproductive function making them well equipped for reproduction (Rufer, 2015).

2.8.3 Genitalia

The organs that are utilized in mating there is the deposition of eggs and all this is known as the external genitalia. Though they are said to be external genitalia these are more so internal. The components' external genitalia of insects are diverse in form and usually have a considerable taxonomic value more so among species that seem to be structurally similar in other respects. In the male external genitalia, they have been used in the distinguishing of species while the female external genitalia that hassimpler and isless varied. The Terminalia of adult female insects involve the internal structures to facilitate the receiving of the male copulatory organ and in the spermatozoa and the external structures that have been used for oviposition. In most cases with regards to the females as they have an egg-laying tube which may not necessarily be present in most of the insects.

The ovipositors usually take two forms which include the true formed from appendages of abdominal segments 8 and 9 or they are a substitution that contains the extensible posterior abdominal segments. These organs are utilized by the males and the females in the case of reproduction allowing them to be reproduce leading to the production of eggs. It is important to note that even for the aquatic insects the process of reproduction occurs on land. This is because the terrestrial environment allows the use of sounds and vibrations among other modes to attract and communicate with their mates (Heckman, 2008).

When discussing the aquatic insects, it is important to note that some insects live in salty water though they are relatively few. Some of the insects that live in salty water tend to look like water strider known as Halobates or sea scatters that allows them to surf the water waves. It has been noted that for the aquatic insects, they tend to live in shallow waters near the shorelines where the light is visible from the bottom. They live in a variety of areas such as mountain streams and rivers. Most of them may dwell in areas that are rich in nutrients. In such areas, there are a variety of insects.

2.9 ECOLOGICAL ROLES FOR AQUATIC INSECTS

For insects, as they dwell in water, there are certain ways in which the insects benefit the ecosystem. For instance, the insects while in the larval stage they facilitate the filtering of water and thereby allowing it to decompose organically. The insects are usually at the bottom of the food chain and thereby allow a variety of aquatic life to survive as the aquatic insects tend to provide food for them. The adult insects do feed on the younger insects. The presence of aquatic insects is an indicator that the water environment is healthy (Heckman, 2008).

2.10 INSECTS LIVING PARTLY IN WATER

In the case of insects, some live partly underwater. For these insects, they live the larval stages underwater more so in the nymph stage. The development occurs underwater. While in the pupae stage they leave the water. They can then fly away as adults. While in the pupae and adult stage they are able to feed on a variety of environments thereby reducing the competition between them. This can be indicated by the adult dragonflies whereby in the adult stage they become land predators and they do not feed on the same prey that they would feed on while they are in the development stages in the water. In the case of insects such as the mayflies and the dobsonflies tend to accumulate energy while in the larval stages. When they reach the adult stage they do not eat the same thing. They feed on what they reproduce while on land.

2.11 INSECTS THAT LIVE ALL THEIR LIVES IN WATER

Some insects totally live underwater. This means that all the stages of digestion do occur underwater implying that the eggs, larvae and the pupae develop in the water. Some live in the mud along the shorelines. For these insects, the adults get to retain the aquatic lifestyle. They can be active both above water and also underwater. In search of new resources, these insects move from one water body to another. Their bodies have been specialized to allow this mode of living. There are orders in the aquatic insects which include the collembolan, Ephemeroptera, Odonata, Plecoptera, Hemiptera, Megaloptera, Neuroptera, Trichoptera, Lepidoptera, Coleoptera, Hymenoptera and Diptera.

2.12 ADOPTIONS FOR AQUATIC INSECTS

Some of the other adaptations of the aquatic insects include the ripple effect which includes the signals being produced by the vertical oscillations of the legs that are in contact with the water surface. The ripple signals occur in the contexts of calling and also courtship, copulation, post copulation, sex discrimination, induction of oviposition, individual spacing, territoriality and the food defense. The other adaptation includes double vision which is in the case of the whirligig beetle that haseyes divided vertically to allow them to see both under and above water. Some have oars which allow them to paddle while underwater through the use of oar-like legs.

The legs have been modified which includes them being long, flattened and fringed. This is applicable tothe case of the water beetle and the water boatman. Some of them have skates that allow the insects to beable to skate on the surface of the water by distributing their body weight over long, thin waterproof legs. This allows them to paddle in the middle pair of legs, steer with the hind legs and they utilize their short front legs to attack and hold prey. Some have jet skis used by the Camphor Beetle when they skate on the surface of the water. When they are alarmed, they release a chemical from their back legs that minimize the water surface tension. There are suckers which are insects present near the fast-flowing streams where the larvae live.

3

Marine Insects

CONTENTS

3.1 INTRODUCTION

Marine insects are known to be insects that spent most of their lifetime in sea or ocean surroundings. It might be inside the sea or outside it. For most marine insects, they tend to survive in the sea and only come out for a while either for fresh air or something else of much importance to them.

The sea is known to be saline in nature. The salinity levels do differ from one sea area to another. This means that marine insects are divided depending on the salinity levels. This is because one marine insect species may survive in a low salinity percentage in seawater as the other survives in a high saline percentage.

The marine insect habitats are known to have been divided mostly into different zones depending on the salinity levels or maybe sometimes the extent to which a certain ocean section is able to receive enough sunlight.

The most common marine habitat is the intertidal (also known as the littoral) zone. They are known to be home to a large number of different insect species. The intertidal zone is further divided into different categories depending on the type of vegetation known to grow in those areas. This is because some insects are known to feed on vegetation and the vegetation on which they feed on is different from one insect to another. The other marine habitat is the supralittoral, sublittoral and pelagic zones. The pelagic zone is known for its host insects which reside in the open oceans which are way too far from the seashore.

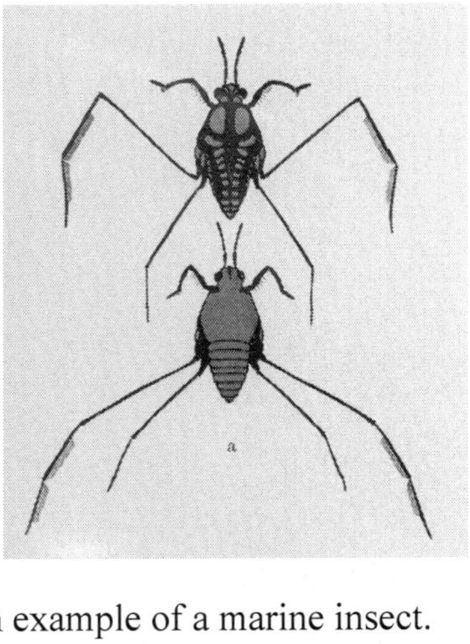

Figure 3.1: Showing an example of a marine insect.

Source: https://www.researchgate.net/figure/Galapagos-Halobates-species-Beebe-1924-plate4a_fig6_309027460

3.2 HISTORY OF MARINE INSECTS

Marine insects are known to be the most evolved category of insects. This is because their habitats have evolved a lot over the ages due to the ever-changing climatic conditions of our planet earth. As we all know, oceans are the most affected by the slightest change in the climate of the earth since they either rise in terms of the sea level or fall in terms of their sea -level. Such a change does affects the survival of insects. Some marine insects are known to evolve to be able to survive in high water levels. Some develop longer siphoning structures for them to be able to harness enough oxygen whenever they float on water. Some develop scales to be able to release the excess salinity that is absorbed by their bodies. In times of fall of the sea level heights, some are known to move closer to the seashore and this means that they have to develop structures that will help them in their survival in shallow waters (Gibson, Atkinson, & Gordon, 2004).

During summer periods, food supply for most insects is quite minimal. The sea waters are known to have fallen in terms of their height. The sea is also known to be in a good supply of foods in all seasons. Such conditions force most non-marine insects to opt for sea life. For them to thrive well in seawater, they have to evolve so that they can survive in saline water and be able to adapt to feeding on aquatic vegetation. This means that the marine insects have increased in number since the creation of the universe and most have evolved greatly to be able to survive quite well in their ever- changing environment.

Most marine insects are known to avoid drowning in water by use of special features on their body surface. Insects can drown even in a droplet of water most marine insects have a waxy layer covering their outer skin which prevents them from drowning in water and enables them to be able to survive in such an environment. The waxy layer repels the water from their skin so that they do not get trapped by the strong forces found in water commonly known as surface tension. Surface tension is known to have a force the water molecules strongly together such that any interference to the tension is prevented by being trapped in the water layers.

3.3 CHALLENGES OF MARINE INSECTS

The survival of marine insects is challenging since the insects have to find ways to survive in the saline sea and ocean water. Some might not be able to survive past larvae level, and some might make it to adult growth level and be able to reproduce too.

One of the major challenges facing marine insects is the oxygen problem (Gibson, Atkinson, & Gordon, 2004). Oxygen is known to cover twenty percent of the atmosphere since the density of air is not that high. The density of water is known to be quite high making water to be quite heavy. Seawater is saline making it heavier than the normal freshwater in most regions and warmer in nature due to its high density. Scientifically, the concentration of oxygen in warm waters and more so waters with high density is quite low. This makes it hard for marine insects to extract enough oxygen from the marine waters as they would normally extract from the air. Most marine insects have developed gills but they do not help that match. This is the reason why most adult marine insects opt to survive partially in water and partially above the water for them to be able to absorb enough oxygen for their survival. In cold sea waters, marine insects are known to survive quite normally like the rest of the insects for they do face that many challenges in extracting oxygen from the water for it is cold and the water density does reduce in with the fall of water temperatures.

Most marine insects are known to face the challenge of predators. More so the insects which have habitats in open oceans. Dangerous animals that rely on insects for food are known to spend most of their lifetime in open oceans preventing the marine insects from floating on the water surfaces for the tapping of enough oxygen for their survival deep down in the ocean water. This challenge forces most marine insects to spend their lives in shallow waters which cannot harbor insect predators.

Breeding grounds of marine insects are known to be invaded by gigantic animal species in open oceans. The destruction of grounds prevents the marine insects from reproducing exposing them to the danger of extinction. The temperatures of open oceans are known to vary from time to time changing the temperatures of the waters. Breeding sites for most insects need to be in warm climatic conditions and with the fluctuating temperatures, it is almost impossible for the young ones of insects to pull through to adulthood or their final stages in such environment (Wehrtmann & Cortés, 2008).

The challenge of the seawater being too salty for insects to survive has been opposed by many for decades some marine insects have been able to survive in such conditions. However, there is some truth in this challenge. Most marine insects may find it way too hard to regulate the salt and water levels in their bodies. With time, most marine insects have been able to adjust to the salty conditions which have enabled them to balance between the amounts of salt and water in their body systems. However, most marine insects are known to not survive past reproduction due to the accumulation of high salt content in their system.

As stated earlier, most insects face the challenge of ending up to be the prey of their aquatic predators. We all know that marine insects need to resurface in the water from time to time for them to harness enough oxygen from the air. Due to predators that are ever alert for the resurfacing insects, most insects choose to move to newer grounds. Not all insects move to new water habitats for this means they will have to take time to adapt to their new environment and the adaptation process might lead to the extinction of some of the insects, hence they choose to remain in their original habitats. Choosing to remain in their original habitats means that they will have to spend most of their time in the deep levels of the sea which is a challenge to them. The respiratory system which holds harnessed oxygen for their respiration is faced with the challenge of collapsing due to the high pressure from the large volumes of water above the insects in the deep areas of the sea compared to the low pressure of the air accumulated in the respiratory system. This might lead to the extinction of some marine insects which are unable to move to new locations.

As repeatedly said previously, the chances of survival of marine insects in the open ocean is quite minimal. This is because the ability to float off all air -filled animal species, more so insects since they store large volumes of oxygen in their respiratory system is quite high. This means that insects will be more buoyant in oceans which poses them to the danger of being prey to the many predators who are always ready to feed on the floating insects. Such risks force most marine insects to move to shallow seawater which is a situation that might be a bit challenging to them since they have to adapt to the new environment and this might take time (Wehrtmann & Cortés, 2008).

The survival of most marine insects is dependent on their ability to cope with the conditions of deep- sea levels for at the surface they face the risk of being prey to the ever -alert predators. The only challenge facing the survival of insects in deep ocean waters is the pack of enough space for them to call home. Crustaceans, which are close brothers to maybe sisters to marine insects, are said to occupy almost three- quarters of the sea beds leaving behind minimal space for the shelter or source of food for marine insects. Furthermore, Crustaceans are known to be well adapted to the conditions of deep -sea grounds and in the competition for survival resources with marine insects, crustaceans win in most cases.

The other challenge of facing marine insects is that most insects end up being said to be of the marine type, breed on land and later move to seas and oceans for their other phase of life. Most of them find it hard to survive in oceans for originally, they were not born with the necessary structures which will help in their survival in salty and warm waters. The migrated insects will face the challenge of navigation around the sea in search of food for growth purposes and mates for reproduction roles hence a decrease in the total number of marine insects.

All the discussed above challenges have ended up limiting the number of marine insects in marine areas. However, quite a few of the marine insects have been able to defy the odds and survived even have more generations to preserve their existence in the face of our planet the earth (Wehrtmann & Cortés, 2008).

3.4 SURVIVAL MECHANISMS OF MARINE INSECTS

Due to the many challenges the marine insects face in their day to day lives as they strive to survive in marine waters, most of them have developed ways in which they can defy the odds and make it to their final stages of life.

In terms of escaping their predators, some marine insects have developed the ability to be able to detect the water ripples which helps them monitor the closeness of their predators. For the marine insects that lack the listening ability, most of them are known to develop their own ripples and be able to observe the echo to detect and predator close to them (Wehrtmann & Cortés, 2008).

The ripples mechanism has also helped insects in hunting their prey. We all know that there are insects that feed on fellow insects. So with the ripple technique, the insects are able to sense the regions in which their preys are available.

Just like any other living organism, insects need to reproduce for them to preserve their existence on planet earth. Marine insects do use the ripple mechanism to communicate with each on their survival ways and also to detect mates for reproduction purposes. This is because not all insects of the opposite gender will reside in the same location at the same time. Some might need to be in different locations for their survival.

The compound eyes of most marine insects are divided in two ways, that is, both horizontally and vertically. This is because a predator, prey, mate or anything else of much importance to marine insects might be located either above the insects' water level, or even below the insects' water level. With such specialized eyes, marine insects are ever alert for anything and always ready and willing to defy the odds for their survival (Del-Claro & Guillermo, 2019).

In terms of their navigation in or on water, marine insects have well-modified legs. Most of their kegs are shaped in an oar-like shape which enables them to paddle underwater. The legs are quite long to be able to move large volumes of water, they are flattened and quite bend in shape and also well spread to reduce the resistance of movement by the large volumes of ocean water. Their bodies in general are well streamlined to reduce the resistance of navigation from one location to another by water.

Some insects have been modified to an extent that they have a special chemical content which is enclosed in their leg structure more so in the pretarsus leg segment. Once the insects detect any kind of danger, it releases the chemical content which contaminates the water reducing the water surface tension at its rare area. The strong surface tension on the anterior part of the insect pulls it at fast rates away from the predator saving it yet another day to enjoy the beautiful life on planet earth and giving it another chance to fulfill its purpose in life.

Most marine insects have developed the ability to balance the salt and water concentration in their bodies. A special physiological mechanism, the osmosis, helps in the regulation of the salt content and water content through different membranes in the body of an insect (Del-Claro & Guillermo, 2019).

As stated earlier, the extraction of oxygen from the dense salty and warm ocean water is quite tricky to most marine insects. Most of them are known to have developed gills and other special structures that aid in the absorption of enough oxygen amount which is fit for their survival from time to time.

Marine Insects have a waxy skin membrane that prevents drowning in water. The waxy membrane also helps them survive in the ever -fluctuating sea temperatures for they do not have to move to warm regions when the ocean temperatures are low. They still do not have to move to temperate water regions when the ocean temperatures are extremely high. In times of strong waves, the waxy layer allows them to stick on the rocky surfaces of the ocean waters and prevents them from being carried away to new locations by the strong sea waves.

All insects are known to migrate every now and then for different reasons. Some insects migrate to escape their predators, others migrate to regions that support their survival, some migrate due to their overpopulation in one region and reduction of competition for different resources. Others migrate just because they located an unoccupied habitat which might perfectly support their survival among other reasons. All these migration purposes are necessary for the survival of all insects despite their habitat. Most insects are known to migrate prior to their reproduction phase in search of a perfect breeding ground. Since all insects must migrate in their lifetime, marine insects have been able to develop special structures and mechanisms which help them in their migration from one place to another. These mechanisms ensure that they safely move from one location to another without having to face the danger of either drowning or ending up being prey to some of their predators among other risks that marine insects face in their marine environment (Rufer, 2015).

For reproduction purposes, marine insects have special features that distinguish them. For some marine insect species, the females might be larger in size and more attractive with beautiful colors compared to the male ones. This distinction helps them during their search for mates for purposes of reproduction hence preserving their existence.

3.5 LIFE CYCLE OF MARINE INSECTS

Insects are known to go through a certain life cycle for them to develop from one stage of their lives to another. This is because insects are covered with a hardened outer layer, the exoskeleton, which does not change its size by either expanding or contracting throughout the lifetime of the insect. For insects to grow, they have to shed the exoskeleton and develop a Jew exoskeleton that will be able to occupy their new body size. The trick in shedding and development of a new exoskeleton is that the exoskeleton will be able to expand to the new insect size before it becomes hardened. This expansion gives a marine insect humble time to grow and fit in the new larger exoskeleton before it time again for it to shed the exoskeleton.

The process of shedding and development of a new exoskeleton in all insects, including the marine insects is known as molting but its scientific name is the ecdysis process. The period in which an insect takes for it to fill the size of its new exoskeleton and have the need of molting is known as the instar period. Once the adult stage of an insect is reached, the molting process ceases for the insects that have already hit its last growth stage and there are no more growth activities that can take place in it. The growth stage of an insect from one stage to another is generally known as insect metamorphosis since the insect does change its general form completely to take up a new form.

Since marine insects are of different species, it means that they will have varying growth phases. Some will undergo two to three phases before they reach their maturity or adulthood levels as others undergo four phases to reach adulthood.

The most common types of insect metamorphosis are two. We have the incomplete insect metamorphosis and the complete insect metamorphosis.

The incomplete metamorphosis, as the name suggests is known to be experienced mostly by the less evolved marine insects. This is because they do not have to undergo any serious changes for them to be termed as perfectly fit for their new environments. In the incomplete metamorphosis, the insect undergoes basically through three phases in its lifetime. The egg phase is when the female adult insect lays the eggs. From the egg phase, we go to the nymph phase. The nymph is said to resemble a tiny adult marine insect but it cannot reproduce, for example, it is a female marine nymph, it cannot lay eggs. From the nymph phase, the insects molt into the adults' phase which is the final phase for marine insects that undergo incomplete metamorphosis. The nymph is known to feed on the same food type as the adult insects. This causes competition between the two phases. Since the incomplete metamorphosis does not allow an insect to undergo all the four stages of development as in most insects, they skip one step, which is the pupa phase. Insects in this phase are quite vulnerable for any challenge they face might leave them completely destroyed. This is an advantage to marine insects that have incomplete metamorphosis for they are at least assured of having to grow to their adult level. During the nymph instar, most adult physical features develop so as to move the insect to its last molting stage to adulthood. The known marine insect species known to undergo incomplete metamorphosis are the true bugs among others.

Complete metamorphosis is experienced by the most highly evolved insect species on the planet. And as we all know, marine insects are said to be the most evolved insects since most of them moved from land into sea in search of new breeding grounds, food among other needs. The complete metamorphosis has four stages through which an insect must go through before it becomes an adult. The first phase is the egg phase which is laid by an adult female insect. The second phase is the larva phase which is also known as the caterpillar phase for most insects in this phase take the characteristics of a caterpillar. It is the growth phase unlike in incomplete metamorphosis where the nymph phase is the growth instar. The larva is said to be completely different from the adult for they do have the same physical or structural or behavioral characteristics, and are known to rely on different food sources. This difference is an advantage to insects since the different phases do not have to compete for food meaning more individuals will be fed. The third phase is the pupa commonly known as the chrysalis. Pupa insta is known to be the period where an insect reorganizes itself in terms of its internal structures with no changes being visible from the outside. The insect is said to be at rest since the internal organs have to be broken down to form a soup- like solution which is used as food by several insect growth buds that end up forming the adult insect. Since the insect is at rest, the pupa stage can also be referred to as the resting stage of an insect. Once the reorganization process is complete, the adult insect is now ready to be formed. The insect moves from the pupa stage to the adult stage under suitable conditions. In the adult phase, which is the last phase of complete metamorphosis, wings are known to develop and be modified to help the insect in its navigation in seawater or other feeding grounds. Marine insects known to undergo the complete metamorphosis are the beetles, moths among other insects.

Not all marine insects are known to develop either through the incomplete metamorphosis or the complete metamorphosis, some are known to hatch directly from the egg and never to develop, that is, they remain to have the same body size since the exoskeleton is known to never expand.

3.6 CLASSIFICATION OF MARINE INSECTS

Marine insects are known to be of different species. Marine biologists made it their mission to study the different marine insect species for further understanding ofhow they impact human life. Several taxonomic groups were discovered. Each taxonomic group is said to deal with the marine insects which have almost similar characteristics.

The first taxonomic group for insects is the Apterygota order. The Apterygota order is further divided into different families some of which are known to be land insects and others are known to be marine insects. Of all the Apterygota families, the Collembola family, the Microcoryphia family and the Tysanura family are the families known to comprise of marine insects. The families are further divided into different genus. Petrobius genus comprises of marine insects which set their habitats under the ocean rocks. The Collembola family is made up of the following genus; the Onychiuridae, Hypogastrudiae, the Neanuridae, the Isotomudae, the Entomobyridae and the Actaletidae genus which are known to reside in the intertidal marine insects' habitats. The Neanuridae genus is found mostly in the rocky upper intertidal zones of the oceans and in the salt marshes which are found along the coasts of many oceans in the world. This genus is known for its unique characteristics of being able to remain submerged under the seawater clinging on the rocks in high tides to protect itself from distraction by the strong water forces which are associated with high tides. They protect themselves by surrounding the body surface with an air bubble that acts as the compressible gas gill which provides it with enough oxygen till the high tide period is over. In times of low tides, marine families belonging to the Apterygota order are known to always move away from light penetrating deep into the sea. However, immediately after high tides, these marine families are observed to be moving towards the sea areas with enough light from the sun. When their population increase in numbers, these marine insects are known to let themselves free and not cling on the rocks in high tides allowing the strong water forces to move them to new grounds in which they can set out their habitats.

Heteroptera taxonomic group is known to be the most important marine order of insects. It has nine families in total which are purely made up of marine insects. Among the nine families, four families are known to be semiaquatic in that they spend half of their time submerged deep into the ocean water and the other half just floating on the water surface. The other five families are said to reside at the shore of most oceans and only one family among the five is purely aquatic, in that it can spend a large part of its lifetime in water without having to wonder on the seashore or float on the water surface.

One of the nine families of Heteroptera order is the Gerridae family. It said to be the most outstanding family in the marine insect environment with it being divided further into three subfamilies. The subfamilies include the Halobatinae subfamily, the Trepobatinae subfamily and the Rhagadotarsinae subfamily. Just like a never -ending chain, the subfamilies are further divided into different genera that include the Asclepios, the Halobates, the Stenobates, the Rheumatometroides and lastly the Rheumatobates all of which have over sixty species each. The Gerridae family is said to be distributed in various marine insects' habitats for it can be found in the nearshore mangrove habitats, in the intertidal reefs, in the coastal lagoons, bays and estuaries, to the open oceans. Of this family, the most studied genus is the Halobates which is known to dominate in the open ocean defying all the odds by overcoming the serious dangers known to face insects in the open ocean habitats.

The second family is the Veliidae family. It is also divided into three subfamilies which include the Rhagoveliinae subfamily, the Microveliinae subfamily and the Haloveliinae subfamily. The subfamilies are further categorized into five genera which comprise of the Trochopus genus known to have five species, the Husseyella genus known to have three species which are confines in coastal bays, mangroves and estuaries, the Xenobates genus which are common in mangrove plants with sixteen species, the Haloveia genus which is said to be the largest genus of the Veliidae family with more than thirty-five species and the Haloveloides genus. In total, the genera are made up of approximately fifty different species.

The third family is the Mesoveliidae family. It is made up of two genera which are part of the marine environment. Most of its species are part of the fresh water habitats with the halophilous and the speovelia species dominating in tidal water which is part of the marine environment hence its categorization in the marine insects' taxonomic groups.

Hermatobatidae family is the fourth family of the Heteroptera order. It is said to be monotypic. There are eight recognized species of this family which are all known to reside in the rocky habits of the ocean where they hide themselves during the high tides and come out during the low tides of the ocean.

The fifth family is the Corixidae family. It is known to be the only true Heteroptera order family that can survive in marine environments. It has around twelve species thatreside in saline environments despite the salinity levels in their environments. One of its species is the Trichocorixa species which is found in salt marshes pools of the sea and the plankton tows close to the seashore. Different from most marine insects, the Corixidae family is said to have wings and actually do navigate from the location of the ocean to another by flight.

The remaining families include the Saldidae family, the Gelastocoridae family, the Ochteridae family, the Omaniidae family and the Aphelocheiridae family all of which survive in marine environments. The Saldidae family is made up of approximately fifteen genera with almost more than fifty different marine species. These families are known to be most dominant in shore bugs. Most of the species in these families are said to be the strong fliers on the sea.

The other marine insect order is the Homoptera order. They are known to be poorly represented in the marine environment meaning they are not that many. In the Aphididae family, the most known species is the Pemphigus species which is known to feeding on the roots of salt marshes along the coastal regions. Other species include the Delphacidae, the Issidae, the Cicadellidae and the Cicadidae, all known to feed and breed in seagrass beds which grows mostly in the tidal salt marshes.

Coleoptera order does comprise of several marine insects. It has more than twenty families that reside in marine environments and depend on them for food and reproduction purposes. Almost all of the families are not purely aquatic for they do spend some of their time on the sea. This order is known to dominate in the intertidal zones of the oceans more so in the sandy areas, rocks, areas dominated by algae which serve as food for various marine insects, salt marshes, among other habitats. The Coleoptera order is made up of different families.

One of the Coleoptera families is the Staphylinidae family. It is the most important family of the Coleoptera for it is made up of more than three hundred marine species. One of the commonly known genera of this family is the Cafius genus which is purely marine and it is made up of approximately fifty species that are known to reside in wrack beaches for purposes of food and breeding. The Bledius genus is said to have approximately four hundred species. This family is not that much of a predatory group. The female insects are known to lay eggs in the burrows and protect them from fungal infection or attack by predators till they hatch into your ones or they molt into the first phase of either the two known insects' metamorphosis.

The Carabidae family is further divided into the Cicindelinae subfamily that contains approximately two thousand species known to dominate terrestrial habitats. The adult insects of the Cicindela species are said to dominate in the intertidal zones of the ocean in low tides and in high tides are found in large numbers in the seashores. Female insects of this family are known to lay eggs which after hatching remain in the larval stage of the complete metamorphosis for almost a year before they molt into adults. The females choose their breeding grounds depending on the salinity levels of the sea bed soil.

The other family of the Coleoptera order is the Curculionidae family which is known to dominate in the driftwood or stranded seaweed along the beaches where they are known to breed and feed.

The other families of the Coleoptera order include the Chrysomelidae family, the Cerambycidae family, and the Tenebrionidae family.

The other of the marine insect is the Trichoptera taxonomic group. Some of the families of this order are not to reside in freshwater environments which eliminates them from the list of marine insects with only a few of its families being part of the marine environment. The marine families include the Chathamidae family and the Philanisus family which are known to be completely part of the marine environment. The two families are further divided into four species with each family being comprised of only two species. Insects in this order are known to feed mostly on the intertidal algae that remain submerged after high tides.

Diptera order is made of insects from different environments. One of the environments happens to be the marine environment. Diptera insects of marine environments are known to dominate the salt marshes. Some of the marine families of Diptera order include the Culicidae family which has the butting midges, the Ceratopogonidae family which has the horse flies and deer flies, and the Tabanidae family which are known to be of great medical use hence being considered of economic importance in most countries. Most adult insects in this order are known to be winged since they spend half of their time in marine environments. The larval stage of the Diptera insects is known to be purely aquatic with most of their time being spend in marine environments. Diptera insects are known mostly for their biting characteristics but there are some of their families thatare non-biting and belong to families such as the Coelopidae family, the Dryomyzidae family, the Muscidae family, the Borboridae family and the Anthomyiidae family. The larvae are known to feed mainly on submerged marine algae. Insects in the pupa stage are known to float on the water surface periodically depending on the light penetration into the sea from the sun, the lunar and the tidal cycles.

The other order is the Orthoptera order. Most of the families of this order are not associated with marine environments. However, the Trigoniidae family has a subfamily known as the Nemobiinae family whose four genera are known to be purely part of the marine environment. The male insects are known to use chemical means communication during the periods they are looking for females for purposes of reproduction.

The Lepidoptera order is known to comprise of purely marine families. One of its species, the Tatoboyts of the genus Pyraustinae of the family Crambidae is said to reside in the intertidal zones of the oceans.

There are several taxonomic groups of insects thatreside in marine environments that have not been studied. Some might be partly studied leaving marine biologists with the challenge of naming them and finding an outstanding distinction between them. Other marine insects have been studied and named but their outstanding characteristics have not been clearly defined. Such orders include the *Anisolabis littorea* order which is also known as the Dermaptera order, the Neuroptera order and the Hymenoptera order which are said to be part of both land insects and marine insects.

3.6.1 The Halobates

As stated earlier, Halobates are a genus known to long to the family of Gerridae of the order of Heteroptera. It is known to be the largest genus of all marine insect genera. Most marine biologists believe it is dominant in open oceans and the only insect that spends all of its lifetime in marine environments. Halobates are also referred to as sea skaters. They said to have approximately over forty species as a genus. Theirspecies are believed to co-exist with other genera of water striders in sheltered marine habitats. However, a larger percentage of Halobates species live in the open ocean area and only manage to come close to the seashore in times of storms that blow them away from the open ocean to the shore. Among all marine insects, the Halobates genus is the only known genus that contains purely aquatic offshore insects, that is, insects which do not have to regularly move to the seashore for breeding or feeding, but remain in the open ocean forever unless forced to move by external forces such as tides, storms, among others. They are known to thrive mostly in marine environment that has tropical and subtropical climatic conditions.

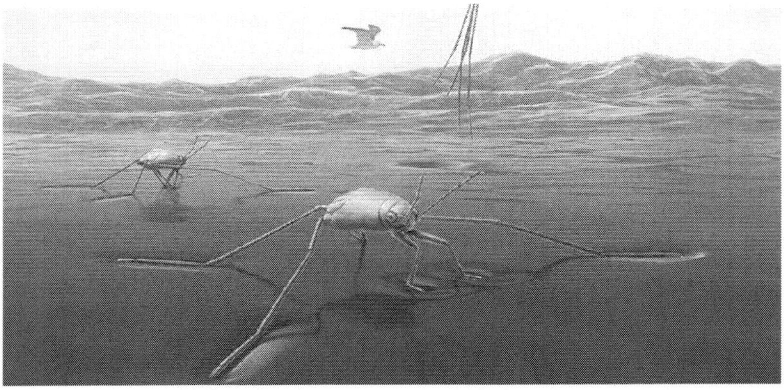

Figure 3.2: A Halobate Insect.

Source:https://phys.org/news/2020-05-sea-skaters-super-source.html

The General Structure of Halobates: All species in the Halobates genus have a general appearance, in that, most of their structures and behavioral patterns are quite similar. When it comes to their structure, Halobates are generally small insects made of a body that is approximately six to seven millimeters long and approximately three millimeters wide. Their leg span is estimated to be at least fifteen millimeters. Like moat marine insects, Halobates are wingless for having wings will not help them that much in their sea life. They have a long antenna to enhance their sensitivity so that they are ever alert on any invasion of their environments by their predators or when they are close to their prey. They have an abdominal segment, mostly the anterior segment, being made up of short legs used for catching prey. In male insects, the short anterior legs are used to hold the female insects during the breeding season for reproduction purposes. The middle segment of the abdomen does have long legs used mainly for the propulsion of insects around their territory. The posterior segment of the abdomen is known to have way shorter rare legs which specialize in steering the insect from one location to another.

Halobates are known to undergo incomplete metamorphosis. They have three life stages for them to termed adult insects. The nymph phase, like in all insects that undergo the incomplete metamorphosis, does resemble a small adult insect. This is a disadvantage for the adult insects that will have to compete for food and other resources with the nymph instar of an insect.

When it comes to the different sexes of Halobates, they do have some close similarities which are quite difficult to distinguish if one is not that keen. The male insects are way thinner than females. The posterior segment of the abdomen, which is the last part of an insect body, has been modified into genitalia, for the male insects, which help during reproduction. As for the females, they are said to have a plump abdomen. The plump abdomen in females is used by male insects mostly as a holding location during the breeding season, for we did say that the short anterior legs of male insects are used for holding the females during breeding.

Distribution of Halobates: As stated earlier, Halobates are known to be dominant mostly in the tropical and subtropical marine environments all over the globe. They are said to thrive well in temperatures ranging from approximately twenty-four to twenty eight degrees Celsius. They cannot survive in waters with a temperature below fifteen degrees Celsius. Most of the coastal species of Halobates are known to stay in small groups in one location, more of cluster families, while other species are known to be all over the coastal regions. Some halobate species are known to occur close to the mangrove plants or other marine plants.

Most of the Halobates species known to dominate in open oceans include the Micans species, the Germanus species, the Sericeus species, the Spendens species, the Sobrinus species among others. The Micans are found mostly in offshore regions which range from forty degrees north to forty degrees south of the latitudes.

Behavior of Halobates: Halobates are known to predators that feed on insects from other environments such as the land which happens to fall accidentally into ocean water. They also do feed on some fish eggs, the larvae of fellow marine insects, the zooplankton vegetation that grows on sea beds and on small jellyfish which are said to be more of sisters to Insects. When it comes to being predators to fish, a tiny fish species may serve as a meal for only one Halobate Insect bug when it comes to a large fish, two or more Halobates come together to feast on the fish till it's done.

Halobates are known to be quite weird insects for they do feed on each other, that is, a nymph can feed on the eggs of a female Halobate, older nymphs may feed on younger nymphs, adult insects may feed on their own nymphs or feed on younger adults. However, the main trick of the Halobates feeding on each other is that they do not feed on their own age, that is, a young insect cannot feed on its fellow young insect. Halobate species are different when it comes to choosing their prey. Some species are known to feed on the immobile prey while others prefer having to struggle with a mobile prey till they get hold of it and feed in it.

Just as the feeding behavior of older nymphs and adult insects is well known, the feeding patterns of newly hatched nymphs remains to be a mystery to many scientists. This has lead to assumptions among marine biologists whereby some say they might feed on the surface film organic components.

Halobates are generally known to be fast in terms of their navigation from one location to another. It is estimated that a healthy adult insect belonging to the Halobate family travels at a speed of one meter per second.

In breeding seasons, some Halobate species are known to Kay their eggs close to the water surface on rocky areas or the plants that tend to float close to the water surface. Other species are known to attach their eggs on floating materials on seawater such as the feathers of sea birds among other objects.

Each female insect is known to lay approximately one to twenty eggs per breeding season. The eggs are said to be white in color, quite transparent and being approximately one millimeter long and half a millimeter wide. The eggs can hatch from anywhere, be it on the water surface or just below the water surface, it has never been a big deal so long as a healthy nymph is produced.

Halobates are faced with the danger of being used as prey by various sea animals. One of the predators to Halobates issaid to be the storm petrels. Storm petrels trick the Halobates by splashing water with their legs which helps them attract sea striders or help in detecting their location. Sea birds especially the noddies do also feed on Halobates. A large number of surfaces feeding fish species are good predators to Halobates.

Walking Mechanisms of Halobates: We did state earlier that Halobates do have short anterior legs which help them in propulsion. The main secret of their navigation is that their legs are covered in tiny water- repellant hairs and on their feet which allow them to tiptoe on the water without facing the risk of having to sink deep into the waters. Still, on avoiding sinking, the hairs do help in balancing the general weight of Halobate insects by spreading the insect's weight over a large surface area on the water. All waters are known to have a pulling force in their molecules which creates surface tension. This tension is known to alter navigation in most insects, however, when it comes to Halobates, the surface tension is never a problem to them, infact, it is an added advantage to them. This is because the interaction of the water surface and air molecules allows Halobates to stand or move on the water quite well. Addition of pollutants such as detergents from water vessels, the ship, reduce the surface tension, and ,in turn, the insects now face the risk of sinking for they try to flop on the surface and eventually end up sinking deep into the ocean. The sinking of Halobates might not be that risky to them, however, if it is unexpected sinking, the Halobate may suffocate due to lack of enough preserved oxygen in their respiratory system.

The insect body of a Halobate is covered in tiny hairs. These hairs are known to trap enough oxygen which helps in generally reducing their weight hence they can float on water. In times when the insects happen to submerge into the seawater accidentally, the air provides enough push for the insect to pop up on the water surface and continue with its life normally. The air so traps enough oxygen which makes the Halobate to be more of enclosed in an air bubble. When the insects sink deep into the seawater for breeding purposes or feeding, the air bubble provides enough oxygen to the insect till it is ready to be back to the water surface to trap more oxygen from the air.

Existence of Only One Genus in Open Ocean: Theories have been derived to try and explain why the Halobates are the only marine insects known to spendtheir entire lifetime without having to migrate to shallow seawater. Marine biologists have questioned themselves over and over again as to why Halobates prefer warm open ocean water, and why they are quite a few, how they manage to survive in an environment which barely occupies other insect genera, among other many questions. Not all answers have been answered, but some have found their guesses which seem to be true in one way or the other. The unanswered questions remain to be a mystery to many, how does it occur that only one genus is able to exist in a vast ecosystem considered to be dangerous by almost all the other insect species.

Several assumptions have been made by various marine biologists in an attempt to bring to an end the mystery of Halobates existence. One of the assumptions is that Halobates are the only insects who have managed to evolve completely to an extentof being able to adapt fully to the extreme salinity levels found in open ocean waters. The second assumption is that, with the feeding habit of Halobates, of having to feed on each other, they have managed to balance their numbers such that they do not have to move in such new habitats. They also have enough room to be able to complete their metamorphosis phases. Their young ones are fully adapted to having to survive in deep waters for quite some time till they hit their maturity level.

It is said that the combination of the depth of waters in open oceans and the high salinity levels in these areas cannot support insect life. Halobates are known to defy this old with the help of their tiny hairs that cover their entire body giving them the ability to harness enough oxygen for their survival in these environments. Most insects that find it hard to be pure marine insects are known to spend half of their lifetime in water, and half of their lifetime in the sea. This has limited their chances of having to evolve completely to fully adapt to the marine environment. With Halobates, they spend their entire lifetime in the marine environment, hence they have a humble time of evolving and developing features that will help them in their survival in such ecosystems.

3.7 HUMAN IMPACT ON MARINE INSECTS

We as humans are slowly destroying the life of marine insects. Human activities that are slowly degrading the value of marine insects include activities such as overfishing, destruction of habitats, the introduction of invasive species in sea waters, ocean pollution, ocean acidification, ocean warming among other activities (Rufer, 2015).

Overfishing, practices are leaving marine insects such as Halobates that feed on small fish with no food to feed on. Lack of enough fish leads to low plankton in the ocean leaving marine insects with little zooplankton to feed in. The predator fish help in maintaining the balance if Halobates by feeding in them so that they do not have to move in search of new habitats due to overpopulation. With less predator fish, the Halobates are increasing in number hence more competition for food, breeding grounds, habitats among other struggles.

Seagrass meadows along the coastlines are known to be a major habitat for several marine insects. With the distraction of these grounds by human beings in an attempt to clear their navigation routes by clearing seagrass and mangrove trees is leaving most marine insects with no place to call home. The coral reefs have been used by marine insects as shelter regions in times of high tides for the insect to attach themselves to the reefs to avoid being carried away to new habitats by the strong storm's waves. Coral reefs are slowly declining in number due to the release of chemical waste into seawater which alters the biological process involves in the formation of coral reefs along the coastline. Marine insectsare being left with no place to hold on to in times of high tides.

In an attempt to increase the sea animals, human beings are destroying the already existing sea life. The introduction of new species in a different environment might be of great significance to marine insects. For instance, the new species might turn out to be predators to marine insects. By the tine insects such as the Halobates adapt to the protection mechanisms against their new predators, large numbers of Halobates will already be destroyed by the new predators. The vice versa is also possible. That is, the new species might turn out to be quite good prey to Halobates, with increased prey, the Halobates will increase in number to hence create an imbalance in the ecosystem (Rufer, 2015).

Marine pollution is slowly leading to the extinction of marine insects. Release of industrial waste into oceans, washing away of agricultural wastes such as fertilizers from the nearby farms into the seawater among pollution means are destroying marine life. The introduced chemicals might alter the biological processes that take place inside and insect's tiny body hence affecting its processes completely. The wind-blowing of debris into seawater might end up blocking the respiratory systems of marine insects leaving them to die. The dust from the debris might also clog the hairs that cover a marine insect's body leaving it with no room for harnessing enough oxygen for its survival in the ocean or its navigation in the sea for the hairs do help an insect to float or tiptoe on water.

An increase of the nutrient content in ocean water leads to a process commonly referred to as eutrophication. Eutrophication is the increased growth of water weeds which end up covering a large water surface. The growth of such water weeds slows down the circulation of oxygen inside the water. Lack of enough oxygen will lead to the death of submerged marine insects for they will not be able to receive enough oxygen from the surface.

The release of toxic chemicals into the oceans are slowly affecting the food chain of most Marin life. The vegetation on which marine insects feed on is concentrated with these toxic chemicals. Once an insect feed on the contaminated vegetation, its biological health is affected fro it has introduced a new substance into its body that is quite harmful.

Plastic pollution has increased in our oceans. The plastics end up floating on water just as normal planktons. Marine insects such as the Halobates genus that is known to feed on zooplankton might not be able to differentiate between plastic and zooplankton and most of the end up feeding on the plastic. Ingestion of plastics into their systems causes suffocation to most marine insects for plastics cannot be digested by any marine life (Rufer, 2015).

Noise pollution brought about by human activities in the sea affect the general livelihood of marine insects. This is because the insects might confuse the insect to be an incoming predator and submerge deep into the sea. If the noise lasts for prolonged durations, marine insects might end up suffocating dye to lack of enough oxygen or the collapse of their respiratory system because if the high pressure by the large volumes of water above them.

The increase of greenhouse gases that have been released into the atmosphere by industries is quite dangerous more so if it has been released by industries located close to oceans. This is because oceans are known to absorb the released carbon IV oxide to form carbonic acid. Carbonic acid is known to increase the acidity of ocean water. Not all marine insects are used to high acidity levels hence increase in acidity might damage their structures which might to their death. Most aquatic plants on which marine insects rely on for food are affected by the increase in acidic levels of the ocean waters. This makes them be quite harmful to most marine insects for they are not really used to them.

Global warming which is also caused by the increase of greenhouse gases in the atmosphere does affect marine insects. Global warming leads to a rise of sea levels due to the melting of the arctic and Antarctic glaciers. Not all marine insects are used to coastal habitats with high water levels, hence this might lead to the death of some of them by the time they adapt to the new water levels. Global warming might also increase the general temperature of seawater. This is also a challenge to marine insects for most of them such as the Halobates are known to thrive well in waters with not more than twenty-three degrees Celsius (Rufer, 2015).

3.7.1 The Impact of Marine Insects on Human Life

Despite their low numbers and quite minimal interactions with human beings for human beings are known to mostly reside on land, marine insects do have an impact on human life. Marine Insects are known to transmit diseases to human beings. Insects such as the butting horse flies, deer flies and midges are known to travel around the coastal lands and cause diseases such as Leishmaniasis to human beings. Most of the impact marine insects have in human life tends to be negative in nature. The positive impacts are quite minimal to an extent that only a few people pay attention to them.

The study of marine insects still remains to be a mystery to many. This is because marine biologists do face various challenges in their attempt of studying marine insects. Some of the challenges include issues such as attack by other marine animals, lack of proper equipment to capture the marine insects in perfect conditions, lack of enough marine biologists who are willing to explore the deep waters of the ocean among others.

Marine insects are quite many. Their impact on ocean life in general, on human life, on agricultural life among other areas is quite enormous. Marine insects life remains to be a mystery. It is high time governments invest in the study of marine insects. This is because apart from the negative effects they are said to cause to human life, many positive effects might also be discovered. For example, there are speculations that some marine insects can be of medicinal value to human beings and this can only be proved to be true if marine insects are studied.

Manual introduction of new marine species in a new habitat should be done after thorough research of what are the positive and negative effects of that introduction. Sea vessels should be modified such that they do cause noise pollution. Industrial or agricultural wastes should be treated before it is released into the ocean waters. Industries should come up with ways of treating their gaseous waste materials to try and reduce the accumulation of greenhouse gases which lead to ocean acidification and ocean warming. The release of toxic chemicals into seawater should be stopped. Nutrient content should be released into seawater in controlled amounts.

All of this should be done in an attempt to preserve marine insects. Marine insects might seem to be worthless to many, but for sure, marine insects are worth the struggles.

4

Cave Insects

CONTENTS

4.1 INTRODUCTION

Insects that live in caves are some of the most prominent and widespread troglofauna (animals that live in caves), all -inclusive of trogloxenes, troglophiles and troglobites. When it comes to their ecological adaptations, these insects have a high significance in their senses; physiological, evolutionary, and ecological.

Figure 4.1: These days, when you enter a cave, one can notice that there is no growth of green plants, and apart from a few insects, such as spiders, there is hardly anything else inside the cave. After walking in a distance, one may note a fly or a moth on the wall, some fungi growing on a log or a different species' spider web.

Source: https://en.wikipedia.org/wiki/Cave_insect

Caves are ecological habitats that are unusually well defined when it comes to their place, time and nature. Actually, it is not to our surprise that many different insects live in caves permanently, especially in places that are much deeper in the cave, and have evolved to a specialized level, allowing them to survive in the extreme conditions.

Troglobites are the true cavernicole species, rather than trogloxenes or troglophiles. Carvenicole insect species living in caves rarely have any adaptation to the movement from cave to cave, and therefore the different communities or species are restricted to their cave systems each or certain caves wholly, and they commonly evolve in their home cave systems respectively. There are however some exceptions in this fact and these are commonly those that have been carried by mobile vertebrate troglophiles or trogloxenes, however, there are some situations where several populations may have undergonean evolution from one mobile troglophile population (Judicaël Obame-Nkoghe et al, 2017).

The majority of the time, caves are geologically short-lived, and therefore, the specialized adaptations are quite young in terms of evolution and have mainly risen up fast just as they did in similar ancestors beginning as similar troglophiles in caves that are separated. Majority of the troglobite insects are Blattodean, Collembolan or Orthopteran, for example, considering they have open -air ancestral species in their nature, it would come with no surprise that the availability of a cave leads to the invasion of this cave by opportunistic troglophiles that may undergo wide distribution, as well as similarly undergo evolution, in separate caves in areas that are different.

For many different ancient insect types, caves have proven to be the last refuge, even though we are no longer found in surrounding regions free in the open. This means that such cave fauna, isa huge representation of relicts. It is important to understand that just because some insects live in particular caves, they have not been there since ancient times.For example, it is obvious that the modern troglobitic Onychophora has not been living in the caves they currently live in since the Carboniferous times, however, and they had recently inhabited new caves and had ended up flourishing through exaptation.

Figure 4.2: Some of the commonly known caves include the Time Machine in Daren Cilau which is the largest passage; Ogof Draenen which is the second-

largest cave running approximately 70km in length; and Ogof Ffynnon Ddu which is the deepest cave in Britain, going around 308 meters deep.

Source: https://www.nationalgeographic.com/science/earth/surface-of-the-earth/caves/

Cave insects become specialized and dependent once they adapt to troglobitic existence on the cavernicolous conditions; they are likely to rapidly succumb when suddenly exposed to the outside world. Troglobites, true cavernicolous species, are inclusive of many insects and animals. Fishes, Mollusca, leeches, Polychaeta, Oligochaeta, and planarians all have troglobite species among them. Some of the species found in troglobitic crustaceans include Syncardida, Isopoda, Decapoda, Copepoda, Cladocera, and Amphipoda. The majority of the troglobite species are predatory, including spiders, Chernetidae, Opiliones, Acari, and Chilopoda (Judicaël Obame-Nkoghe et al, 2017).

Some of the insects that are Troglobitic include Apterygota, for example, certain Collembola and Campodea. Within the beetles' families, there are many in this category, including Silphidae, Leiodidae, Curculionidae, and Carabidae. There are also some troglobitic Orthoptera while some are trogloxenic. The orders Diptera, Trichoptera and Battodea all include troglobites.

4.2 CLASSIFICATION OF CAVE INSECTS

- Troglobite species – these are considered the true cave dwellers, where they exclusively occur in the caves and they are not able to survive outside them. Examples of these insects include some Zygentoma, Diptera and Stenopelmatidae, and many Coleoptera.
- Troglophile species – these can sometimes be found outside the habitat of the cave, however, their life cycle is typically completed in the caves.
- Trogloxene species – these do not have the ability to permanently reside in the caves, however, there have the ability to enter the caves and survive through certain stages of their life cycles within the caves.
- Tettigoniidae – these are unusual and found in South Africa in certain genera for example Cederbergeniana, and these are plant

feeding trogloxenes at night, and stay inside their caves during the day.

4.3 TROGLOFAUNA

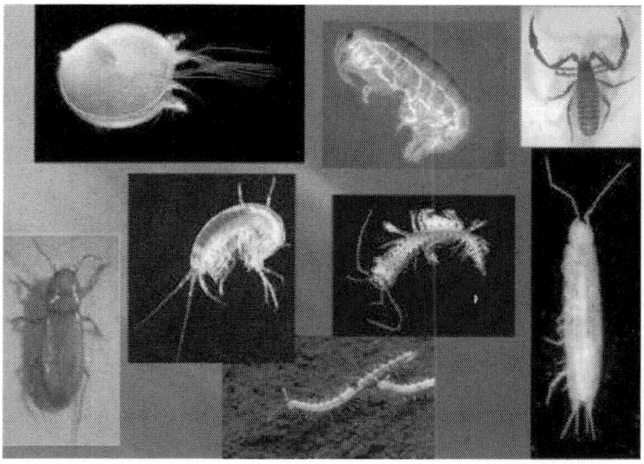

Figure 4.3: Troglofauna is a collective name used to refer to the creatures found and living within caves, and these creatures can undergo classification in accordance to how dependent they are to the environment inside the cave, and how much of their life cycle they spend in the cave.

Source: http://eemb168.blogspot.com/2009/10/troglofauna.html

These are small animals that live in caves and have adapted to dark surroundings. Based on their life history, Stygofauna and Troglofauna are two different forms of subterranean fauna. They are both in association with subterranean environments – in general, troglofauna live in spaces above water tables and caves, while stygofauna are associated with water. Myriapods, insects, spiders and others are all part of Troglofaunal species. Some of these species have to permanently live under the ground and they are not able to survive outside the environment of the cave. The characteristics and adaptations of troglofauna are inclusive of heightened smell, touch and hearing. There is an apparent loss of senses that are underused such as lack of pigmentation and the majority of the troglofauna species lack eyesight. Insects species in this category may lack longer appendages and wings.

Based on the ecology, there are three main divisions of troglofauna:

- Troglobites or troglobionts – these populations of species or species can only survive in subterranean habitats.

- Troglophiles – the insect species in this category mainly live above ground, however, they can also do well in subterranean habitats. We can also further divide these into subtroglophiles (these are more inclined to temporarily or perpetually live underground, however, they have certain functions that they have to do above ground) and eutroglophiles (these species have the ability to live in caves permanently even though they are above ground species.
- Trogloxenes – these species mainly occur sporadically in caves, and they are also not able to establish populations in subterranean environments.

Troglofauna normally inhabit moderate cave regions. There are hardly any changes in the overall climates of the caves all through the year. These caves have high humidity that ranges from 95% to 100%; having low rates of evaporation. We can divide the ecosystem of the troglofauna residence caves into four main zones: the deep cave, transition, twilight and the entrance. Where the underground environment and the surface meet in the entrance zone. In the twilight zone, the light becomes scarce. There is almost complete darkness in the transition zone; however, it is still possible to feel some outside environmental effects. Finally, there is a relative level of stability, no evaporation, and complete darkness in the deep cave zone. The troglobites are normally located in the deep cave zone (Judicaël Obame-Nkoghe et al, 2017).

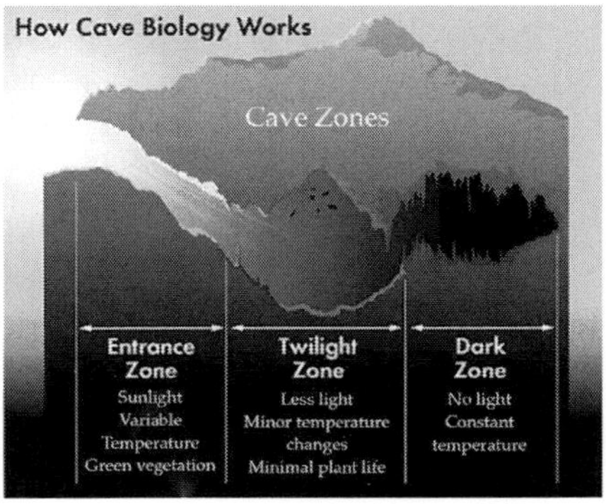

Figure 4.4: Above is a great illustration of the different zones that are found in the cave: the entrance zone, the twilight zone and the dark zone. The three zones are differentiated above.

Source: https://science.howstuffworks.com/life/biology-fields/cave-biology2.htm

These insects have had to adapt to a limited food supply, and they still remain some of the most energy- efficient animals. They find their food from epigean animals such as zooplankton, bacteria, leaves and twigs. They can also find food from feces such as bat guano, egg deposits, and trogloxene carcasses. There are also some predator species such as Troglofaun beetles, which feed on other animals in the same category instead of guano, twigs and bacteria.

A hypothesis on the adaptations of troglofauna was made by Francis G. Howarth when it comes to their ability to exist in the cave environment, concluding that they have ended up losing many of their mechanisms for water conservation of surface relatives, and are now very alike in water balance mechanisms with permanent aquatic arthropods, including cuticular permeability. These insects thrive in environments that are humid and display either comatose or agitated behavior when they are in dry environments, showing that they have a high susceptibility to alterations in humidity and temperature. So as to survive in low oxygen levels and food scarcity, these insects end up developing a low metabolism. Due to this, they have the ability to live much longer than other terrestrial species (Holt, 2012).

These insects have been able to evolve in isolation. Their dispersal is hindered by fluvial barriers such as streams and rivers and stratigraphic barriers such as rock layers and walls. Consequently, their food and habitat availability can provide a wide range of biodiversity globally.

Many animal groups are represented in the Trolglofaunal species, including salamanders, beetles, diplurans, collembolans, isopods, harvestmen, pseudoscorpions, spiders, millipedes, and gastropods. The gastropods in this category are endemic to Europe and the United States; they have a much higher concentration in the northeastern Mediterranean regions. The scorpions are mainly found in caves in Mexico. We are able to find the spiders in Japans, Europe and the United States; however, they may also be found in the Philippines, Australia, Cuba, Congo and Mexico. These animals are found all over the world (Holt, 2012).

There are many caves all over the world (approximately 90%), that have not yet been found because they do not have visible entrances, and there are many other species that live in vugs, fissures, and other spaces that are found above the water-table. In the same manner, there are still many different troglofauna species that have not yet been discovered.

Figure 4.5: The troglofauna are very sensitive to small changes in their environment, especially because they are restricted in them for long periods. Floodwaters are considered one of the major threats to these species, due to the fact that thus, environmental change can lead to a reduction in oxygen and food availability, as well as space for them to move in the cave.

Source: https://en.wikipedia.org/wiki/Troglofauna

Troglofaunal species can be threatened by floodwaters, causing a detrimental effect; the dramatic change in end availability of oxygen, food, habitat, and even connectivity to other habitats. The majority of these species have a high level of sensitivity to alterations in their environment, and floods, which can also be accompanied by lowered temperatures may be life- threatening for many of these animals. Troglofauna species that live near the surface can be affected by extreme winter temperatures. Bats and birds that live in caves feed on troglofauna. This means that the troglofauna end up competing against each other for survival. As humans, we also pose a major threat to these animals. When we mismanage contaminants such as sewage and pesticides, we manage to contaminate the water which poisons troglofaunal communities. We also threaten their existence by removing their habitats, either indirectly (eg rising water-table) or directly, is also a huge threat (Holt, 2012).

4.4 TROGLOBITES

Troglobites are populations of species, or animal species that strictly live in caves and other underground habitats. It is important to note that these are different from species that only visit the caves and species that are able to live underground in addition to on the ground. Troglobites that live on land can be called troglofauna, while those that live in the water can be referred to as stygofauna.

The majority of the troglobitic species have undergone evolution to fully adapt to life within the cave. A few of these adaptations include depigmentation (lacking pigment within their integuments), reduced or full loss of eyesight (anophthalmia), better efficiency when it comes to using food, lowered level of energy consumption, and low metabolism rate (Mills, 2009).

Conversely, despite the fact that they have reduced or lost functions, they also have some that are increased. There are many species that have undergone evolution to attain elongates locomotory and antennal appendages, so as to have the ability to move around better as well as for responding to stimuli in their environment. The appendages also have humidity, tactile and chemical receptors. In general, these species are not able to survive well when they are not in their caves, and this means that they are not able to travel from one cave to another. Due to this, these species are endemic to single cave systems or a single cave.

It is important to note that not all species that live in caves are considered to be troglobites. Animals that are located in caves could either be trogloxene (these only sporadically occur in these environments and they are not able to create subterranean populations), or troglophile (these live both in epigean and subterranean habitats, such as cave swallows and bats).

There has been the discovery of more than 7700 troglobite species. Although this may look like a high number, it is believed to be a very small part of the total existing troglobite species all over the world. This number is considered to be low because there has been a poor exploration of many caves, and there has even been a poorer sense of biological census within these caves. In addition, the number of discovered caves is just but a drop in the ocean of the many other caves that are existent (Mills, 2009).

Figure 4.6: Troglobite species are very unique and cannot complete their life cycle in any other environment other than the inside of the cave. This is so much that they have had to adapt morphologically to the environment inside the cave, for example, they have a lower metabolism, longer legs, and they are blind.

Source: https://www.researchgate.net/figure/Examples-of-subterranean-invertebrates-from-Western-Australia-a-Troglobitic-spider_fig1_236985090

Due to the fact that troglobites evolve in isolation, this also accounts for their high number of different species. One species is able to undergo evolution in one cave, and due to the fact that it is not able to survive in the environment outside that cave, then there cannot be spread of that species to other caves. In general, this explains how each cave can host many different unique troglobite species at once.

There can be an evolution of the same species creatures into different species, in the same cave. For example, if there was wondering of spiders of the same species into different cave passages, they could each have independent instances of evolution in the different passages. This is all because these passages have different unique conditions that have the ability to provide different outcomes in terms of their evolution.

The majority of the species in these categories are sedentary organisms, and they hardly ever burn many calories at a time. Most of the food they eat is obtained from scavenging. Part of what they eat is inclusive of feces of other animals that they find on the floor of the cave, carcasses of dead animals in the cave, plankton and bacteria living in the cave waters, and small particles of plant debris that they collect from the cave floor brought to the cave by running water. Troglobites living in caves that have a large bat population can primarily survive on Bat guano (Judicaël Obame-Nkoghe et al, 2017).

Below we will discuss a few troglobite species.

Neoleptoneta myopica (The tooth cave spider): As of now, this spider is referred to as *Tayshaneta myopica*, rather than its previous name *Neoleptonefa myopica*, and it is 1.7mm long. It is found in the spider family Leptonetidae. This species is mainly found in limestone caves that are found in the United States in Texas. They are considered to be endangered.

In 2013, this species was also discovered in caves in Williamson County and Travis County to Texas in the northwest of Austin. There have been 6 different provided locations for the species. however, the exact detail of the locations has not been provided, mainly so as to ensure that these species are protected. We can find these caves in Central Texas, where there is an occurrence of the caves in Edward's Plateau, a region that is rich in limestone. These spiders end up spending the majority of their time in their webs, all excluding the mature males. Scientists have been able to study members in these species in four different caves (Tight Pit, Gallifer Cave, Root Cave, and Tooth Cave), and all the species have shown to have the same nuclear and mitochondrial DNA haplotypes, which shows that they do not move from site to site. These spiders were located suspending under low sheet webs, where they would drop if they are disturbed.

Figure 4.7: Above is a picture of the tooth cave spider. It is classified as a troglobite because it lives a notably different lifestyle from its counterparts, the Amauropelma, mainly because it lives in the darker recesses of limestone caves.

Source: https://sciworthy.com/cave-spider-defies-tropical-environment-remains-pasty/

*Leptodirus hochenwartii.:*This is a cave beetle that is classified under the family Leiodidae. This is the only species that is found in the genus. This cave beetle is a true troglobite that is found in Croatian, Slovenian and some Italian caves. These beetles are not able to survive in any other environment other than the subterranean environment, due to their adaptations. Due to this, they have features that are considered strictly troglobiotic, such as elongated antennae and legs, they lack wings, anophthalmia (lacking eyes), and also lack of pigment within their integument.

They however have some very interesting features, such as their domed elytrae which completely covers their abdomen which contributes to the beetles' weird round appearance, and their slender throat, which was used to name them (leptos – slender, deiros – neck). In addition to this, they also have a highly specific receptor on their antennae, referred to as the Hamann organ, which provides the beetle with the ability to perceive the humidity of the air.

These beetles mainly live in caves that are both cold and large, where the temperatures do not go above 11°C. They have a largely unknown ecology, however, the available collected specimens were seen to feed on organic materials, both of vegetable and animal origin, all coming from the environment outside the caves through birds' and bats' guano or percolating water, and the carcass of other various cave animals (Judicaël Obame-Nkoghe et al, 2017).

There is even much less that is known about these beetles' life history. There is only one study done on these species (*L.hochenwartii*), as of now, and it shows that in a similar situation with the majority of the specialized Leptodirini, the females lay just a few eggs of relatively big eggs that normally take a long while to develop. There is a reduction of the number of larval instars to only one, and before molting, the larvae do not feed. It is still not known what the maximum period of adult activity is up to this date. As with other insects in this category, these insects have also lost their circadian rhythm due to the absence of sunlight and environmental stability, while the rainfall patterns additionally affect their seasonal rhythm.

4.5 THE CAVE ENVIRONMENT

In situations where the primary molding agent in the cave landscape is dissolution instead of erosion, these landscapers are referred to as karst landscapes. Some of the common features that you find in such landscapes (blind valleys, springs, sinkholes, caves, etc.) are due to rocks being able to hollow out from the effect of weak acids instead of, for example, earthquakes, volcanic activity, or erosion. These landscapes normally account for approximately 16% of the total surface on earth, representing 40% of the United States in Oklahoma, 25% of Italy and France, 45% of Slovenia, and 75% of the total surface area in Cuba.

Caves can be found in rocks that are more than 500 million years old, and also in rocks that are less than 20,000 years old. The majority of the caves we find our created due acidic water action on carbonates (mainly limestones however it may also work with marble and dolomite), as well as evaporites (especially gypsum however it may also work with rock-salt). Many of the large caves are formed using limestone rock, which mainly consists of calcite ($CaCO_3$) as the main mineral. Many other caves form using gypsum ($CaSO_4.H_2O$), which has high water solubility. Caves that are formed from gypsum can run for more than 100 km. However, they are generally much dryer in addition to being short- lived and younger than limestone caves, mainly because they erode 1200 times much faster than limestone caves. In addition, there are relatively very few life forms in gypsum caves (Mills, 2009).

Figure 4.8: With movement from outside into the cave, the nature of plant life changes, and this is correspondent to the habitat changes. There is a notable w the amount of light (decrease), and with an increase in the shelter, comes with a decrease in the wind, thus helping in the maintenance of humidity (in combination with the cave's damp air).

Source: https://www.pinterest.cl/pin/496521927670868006/

The majority of all the caves that are made through this solutional process have several similar components, such as a saturated zone (phreatic zone), streams, and a vadose zone of percolation. The epikarst zone is the one found above the caves and is an area that has cavities, solution tubes, and cracks that are poorly integrated.

Lava tubes are caves that are not formed by the solution. These caves are mainly formed as a byproduct of volcanic processes, and they end up being the same age as the rock. In comparison to solution caves, which can take up to millions of years to form, lava tubes are formed almost instantly. In addition, lava tubes are considered to be quite transient, because they can be eroded quite rapidly or completely covered by flows of the new lava.In Hawaii, the lava tube (Kazumura cave) has more than 66km of passages, approximately 102 entrances, and even a vertical length of more than 1200m, yes it has only been existing for 360-500 years. The majority of the time, lava tubes are shallow and often less than 5.1 m below the surface. Throughout its length, Kazumura Cave is less than 21 m below the surface. It is very rare to find such shallow lines to the kids. There is often the penetration of the lava tube ceilings by the roots of plants on the surface. These root faunas are both diverse and unique (Mills, 2009).

Some other caves occur as a result of tectonic and wave action, in crevices, talus slopes, and ice. Species that have undergone adaptation to live in subterranean life are hardly found in such caves, except in the Brazilian iron ore where enigmatic caves are found, harboring an insect fauna that is rich however largely unexplored.

It is not possible to tell the difference between night and day in the deep caves, and their climate is hardly ever affected by the different seasons passing. They also do not experience violent storms and winds, however, there may be times when there are steady air currents. They have a rather stable humidity, and they only communicate with the world outside the caves in certain special events such as exceptional droughts or floods. In caves that have streams passing or seepage of water, they tend to be the insect's common source of food and water (Mills, 2009).

When it comes to cave ecology, trogloxenes are very important because they have to leave the cave to feed, and while returning import materials that troglobites can use as a source of food. Some insect species such as beetles, flies, and butterflies over-winter in caves, and those that do not survive and die to remain as food. An example of the trogloxenes are Cavernicolous bats, and these have proven to be very important to the ecology of the cave, where they spend the majority of their day when the sun is up, and this provides massive deposits of organic matter, especially the Mexican free-tailed bats, mainly in the form of carcasses and feces.

Other animals are considered to be trogloxenic, including cave swiftlet, oilbirds, reptiles, other predators, hyenas, bears, and even humans, that enter caves for either hibernation or short-term shelter. The majority of these animals help in contributing to the amount of organic matter, instead of consuming it, and have proven to be very important in providing these resources for insects, the majority of which specialize when it comes to relying on certain species that visit regularly for a long time.

It is , therefore, true to say that the cave environment is characterized by attenuation, restriction, or absence of certain factors such as abrupt contrasts in humidity and temperature, freedom of movement, living space, seasonal or circadian stimuli, and light. Other factors may be fully dependent on the conditions that are available locally, for example, there are known the majority of caves that provide very small food quantities, and some also provide small water quantities, whereas others are able to provide great quantities of dung and perennial water that are able to provide for ecological stratification, with the different organisms preying on each other, especially in the various stages of the product that was originally there (Holt, 2012).

The majority of all the food that is found in the cave comes from outside the cave. Carcasses and other organic detritus are carried into the cave by running water and air currents. Bacteria and fungi that develop on these materials carried into the cave provide food for many insects that live inside the cave. Another source of food is bat guano.

Figure 4.9: Conservation of these caves has proven to be a very important act. Caves are a rare and special environment, some being more so than others, and even some specific parts more than others. Cavers have the ability to accidentally damage potholes with fast- moving water and also active streams, which may have a major effect on the cave environment.

Source: https://www.cnn.com/travel/amp/vietnam-hang-son-doong-cave/index.html

Troglobitic Orthoptera prey on Lepidoptera that enters the cave to sleep, the main predators being Gryllidae and Terrigoniidae. Cavernicolous Collombolae feed on dust borne on the surface tension or colloidal matter in the water. Spiders and Myriapoda feed on the smaller insects and other invertebrates. Majority of these activities happen in the dark, except those that happen close to the entrance of the cave, or in situations where bioluminescence is provided by certain insects, for example, Arachnocampa, or microbes, even if it's only for the attraction of prey.

4.6 ECOLOGICAL ROLE OF CAVE INSECTS

There are many important reasons why insects occur in caves and other subterranean habitats. The weird behavior of Heleomyza flies in British caves was described by Chapman in 1993. They end up arriving during autumn and remaining in a state of semi-torpidity for several months, where the majority of them die from pathogenic fungi or starvation. It was suggested by Chapman that seasonal reversal of air currents affect the flies. They are normally able to enter the cave by following air streams that are flowing into the cave, and therefore they end up flying out of the cave by mistake due to the change in airflow. Certain insects end up in the cave with changes in the climate to avoid the winter cold or the heat of the summer. Many of these insects that end up in the caves end up being a source of food rather than residents. Many aquatic animals end up in the caves with water being swept into the cave, and these also end up as a food source (Holt, 2012).

If entrance zone residents and temporary visitors became the main constituents of cave biota, there is a low likelihood that there would be constructions of any special categories, however, a special name (amphibionts) can be used to refer to hyporheic insects, for example, the stonefly *P. frontalis*, that ends up spending majority of its life in subterranean water. The majority of the cave biota, however, is made up of populations that use caves as habitats for the entire duration of their life cycle. The discovery of species that had elongated appendages were depigmented, and were eyeless, brought about the real interest in cave biota. Schiner was able to propose a classification on the ecology, based on habitat, mainly due to the fact that subterranean inhabitants from different groups havethis same early stated morphology, and this classification included a category of those species that could not live outside the caves.

The story of *L. hochenwartii* was figured out in 1851 by Schiödte, who also proposed a classification that was based on the zones of light in the cave, however, as of today this classification is hardly ever used. In addition to the restrictions in leaving the cave and being obligate inhabitants, the majority of these species share these three common characteristics; elongated appendages, eyelessness, and lack of pigment. This system of classification is referred to as the Schiner-Racovita system, and despite the fact that it is based on animal distribution, the major attribute in the majority if not all subterranean animals are the morphology (Holt, 2012).

Of course, the main impetus for including other subterranean habitats that are not caves within the purview of cave biologists and speleobiologists, was the fact that there were many different organisms that had the same morphological syndrome in all of these habitats. The first person to probably comment on the extinction of cave environments was Racovita in 1970, and there was a unity to the study of these environments due to the appearance of cave insects in shallow and interstitial subterranean environments.

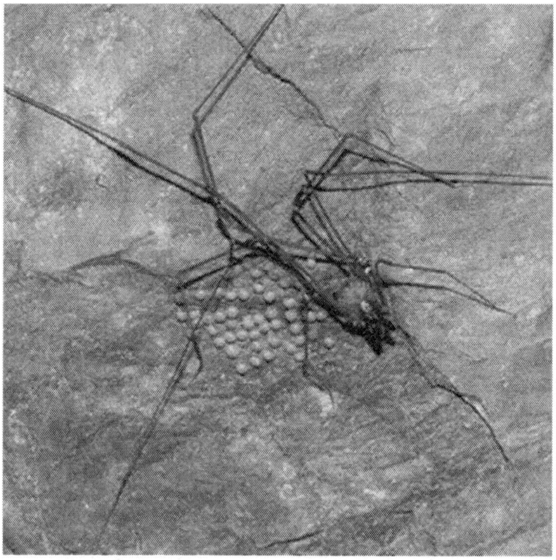

Figure 4.10: Most insects within the cave are found nearer the surface where there is much more food. When going deeper into the cave the harder it gets to find more insects. Much deeper into a cave, there may be insect communities that are concentrated around a certain area. This area may have a source of food that the cave bacteria generates, such as flowstone.

Source:https://www.researchgate.net/publication/314126290_Towards_a_biologically_meaningful_classification_of_subterranean_organisms_a_critical_analysis_of_the_Schiner-Racovitza_system_from_a_historical_perspective_difficulties_of_its_application_and_implica

Troglobites, also referred to as troglobionts as well also utilized to refer to both aquatic and terrestrial species, and the stygobiont which is the parallel category and is used to refer strictly to aquatic species, and is inclusive of all species found only in subterranean habitats. It was recognized that troglobionts and tygobionts were defined ecologically, and in 1962, Christiansen took the opportunity to propose a term to make a reflection of the convergent morphology of animals that are obligate to subterranean environments (troglomorphy). This convergent morphology has the main characteristic of losses (for example cuticular thinning, reduction of both pigments and eyes), as well as gains (for example increases in egg volume, length of the appendages, and extra optic sensory structures (Holt, 2012).

This convergence has helped in making it easier to identify troglomorphy. There is also another fact that is simultaneously both confusing and interesting, and this is the fact that not all troglobionts and stygobionts are actually troglomorphic. Recently, the species thathave been isolated in these habitats may have had little to no time for the evolution of troglomprphic features, or their subterranean environments may have not achieved these specific pressures that lead to troglomorphy. The usefulness and clarity of ecological classification systems help in breaking down the insects especially when there is movement beyond troglobionts and stygobionts. Many other different species live in these cave environments. Some of these include both cave-dwelling and surface-dwelling populations that permanently live in one cave or the other. An example of this includes *Pseudosinella violenta*, which is found in almost all the caves in Texas, and also on surface habitats in almost all parts of the United States. This is a eutroglophile; one kind of troglophile in the system we refer to as the Schiner-Racovita system. Cave-roosting bats, some cave crickets and stoneflies in hyporheic habitats are examples of other troglophilic species that are bound to the surface for specific biological functions, such as reproduction and feeding.

It is ,therefore, true to state that the term troglophile can be used to describe many different species and may also be thought of as a way to define species that are a significant part of the cave ecosystem, however, they are not trglobiotic (or stygobiotic). In the Schiner-Racovita system, the final term to be discussed is the trogloxene, used to refer to the occasional or accidental subterranean visitors. However, this term has been defined in many different ways by different authors, leading to what is referred to as a terminological mess (Holt, 2012).

Within many different cave habitats, including Trophic RoleIn caves, there are many different trophic roles that the insects assume, including as parasites (especially on the bodies of the cave bats), as predators, and as detritivores (mainly because the majority of the cave communities are based on detritus). In North America, the cave crickets have a very unique trophic role. They are not only responsible for bringing organic matter into the caves, but they are also responsible for laying eggs in sandy substrates within the caves. This means, that for a number of carabid beetle species, cricket eggs end up being their main dietary item. It is also possible for cave insects to compete. In 2013, it was discovered by Vergnon et. al. that the concept of limited similarity, which states that due to interspecific competition, species must be different in a sufficient manner for them to peacefully live with each other. The study showed that in 35 calcrete aquifers, having either three or four species of dytiscid water beetles, there was a remarkably constant ratio of their body sizes, with the ratio approximately coming to 1.6. There was a lot more interest in this phenomenon, mainly because there was a variation in the absolute sizes of the beetles among the calcrete aquifers.

4.7 INSECTS COMMONLY FOUND IN THE CAVE

There are many different cave insects, including some troglobites. These will be discussed below.

Springtails: These are in the Collembola group and they are a group of arthropods from the ancient times because there were findings of similar forms from 500 million years ago. These insects have no wings and they dwell in the ground due to their requirement for high humidity. There are many of these species that have been found in Welsh caves. Their name is given to them due to their ability to leap around with the help of a spring or trigger. Underground and soil species have shown to have much smaller springs, shorter antennae and legs, loss of pigment, and reduction in their eyes. These insects feed on animal material, decomposing plant, bacterial and fungal mycelia.

Figure 4.11: Springtails are small creatures that love moisture. They have no wings and look for damp areas both on the ground and underground. The majority of the time, they cluster up in swarms, and when they are disturbed, they jump into the air several inches high, where they are able to create clouds that are easily visible.

Source: https://www.planetnatural.com/pest-problem-solver/houseplant-pests/springtail-control/

Majority of these insects are found on pool surfaces, some being quite tiny (1mm), looking like a spec of dirt, however, when observed for a longer time (about one minute), the can be seen to hop a couple of inches, proving to you that they are living creatures. They are so small and light, that they have the ability to move about on top of the water's film surface. They can easily be located on damp mud banks and still pools. Despite the fact that they hop, not all of them are willing to jump. Some of them make a grazing motion as if walking. (Judicaël Obame-Nkoghe et al, 2017)

One of the common species is the Tomocera minor. This species is around 4.6mm long. It is commonly found on mid banks and uses fungi as its main source of food. Normally, it lives in the soil, however, during the cold weather, it moves downwards. Another species is the Folsomia species, and these are found where the surface isnot high, or near the entrance. All over the world, *Folsomia candida* can survive underwater with low oxygen levels. The Oligaphorura is a sluggish, rod- shaped, white and blind species which can be found deeper underground, and might also be seen on the surfaces of pools with the Isotoma species which is much more active.

One of the true troglophilic species is *Oligaphorura schoetti*, and the collected information on them is all from caves where they can commonly be found. There are also some smaller jumping, globular, pink *Arrhopalites pygmaes*, and these are also found on the surfaces of pools, however, they are not found together with *O. schoetti*. *Pariosotoma notabilis* is a grey -brown species that is approximately 1.4mm long and is found deeper underground. In the UK, the *Schaefferia emucronata* is commonly found in the caves, however, its anatomy has not yet been fully understood.

Diplura: These are primitive insects that lack eyes and have segmented antennae. They live under leaf littler and underground in the soil. They have a resemblance to white earwigs and feed on dead organic matter and prey on springtails and fungi. They have a usual characteristic that involves them regenerating body parts that they had previously lost. On certain species, they use their tails as pincers.

Figure 4.12: Diplurans are normally located on almost all of the landmasses, apart from Antarctica as well as certain oceanic islands. Their purpose as organisms that dwell in the soil and caves may help in indicating the quality of soil, as well as for measuring the anthropogenic impact.

Source: https://en.wikipedia.org/wiki/Diplura

Diptera: This order includes insects that have undergone modification of their hind wings into a pair of balanced organs referred to as halteres. There are many different Diptera species found within the caves, however, the majority of them are only in the caves as visitors. It is common to find these flies in the same place every single year. They have two main sub-orders: the Brachycera and Nematocera. These will be discussed in detail below.

- Sub-order Nematocera: These are also referred to as 'thread horns', and one can notice them due to their feathery antennae and elongated bodies as represented by crane flies, midges, gnats and mosquitoes. Majority of these swarm. These families are described below:
  - *Family Mycetophilidae (fungus gnats)* – their name is derived from the fact that they feed on fungi. These insects are highly delicate having long legs (especially the hind ones) and long, thin antennae. Insects in this family, for example,*Speolepta leptogaster* species which is found in many different caves, are considered to be troglobites, mainly because researchers have not yet been able to find their larvae outside the habitats of the cave. The adults have a certain resemblance to crane flies with an olive- brown color and long legs. The larva is brown or yellow -headed and is translucent, living in a web of thin threads that is similar to the sheet of spider webs, feeding on fungi and bacteria. This thread entanglement is mainly to make sure that the insects do not touch the walls, preventing them from becoming wet and susceptible to fungi. However, this entanglement of threads could also be used to trap organic material and fungi because the larvae feed on the web together with anything trapped in it. These strands of the web can easily be found on the wall or against the roof, many times with the larvae moving on it. It is very hard to find the adult because they fly a lot from one point to another, and also because they probably do not live long.

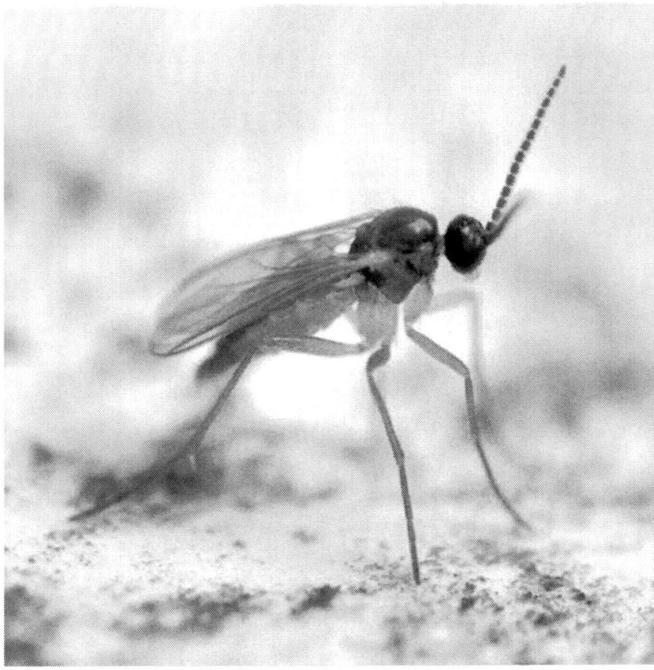

Figure 4.13: Despite the fact that fungus gnats occur in high species numbers, in moss-covered rock areas, cavities between tree roots, Hohlwegen, shady forest roads, bush-lined spots and torrents, it is actually quite hard to see them with the bare eye, either because they are quite small, or their color is hard to see.

Source: https://www.planetnatural.com/pest-problem-solver/houseplant-pests/fungus-gnat-control/

- *Family Sciaridae (Sciarid flies) (dark-winged fungus gnats)* – an example of this is the Bradysia species that has smokey wings and a black body (mainly for the males), where their adults are much more commonly found that the larvae, unlike the species described above (Speolepta). There is not much that is known about these flies. The adults are quite similar to mosquitoes, with large eyes and relatively long wings and legs, with beaded antennae, however, they are weak flyers. Their larvae aresemi- transparent, black -headed, and have12 segments. The larvae spin threads for creating cocoons and they also have the ability to create much simpler webs. The larvae use fungi as their source of food, while the adults feed on animal material and decaying plants. When the adults die, their remains are normally covered in fungi.

- *Family Trichoceridae (Crane Flies)* – this isreferred to as the crane fly or the winter gnat *Trichocera maculipennis*. They are approximately 7mm long and can be found in large cave systems deep underground, for example in Ogof Ffynnon Ddu and Agen Allwedd together with their yellow larvae. There is a wide distribution of the adult flies, however, they are quite hard to spot. Rotting meat can be used to attract the larvae as bait. They look like crane flies, however much smaller and delicate. During the summer, other craneflies visit the threshold, and these never feed as adults. In cave thresholds, the species *Limonia nibeculosa* visits only during the summer. Unlike many crane flies, it rests with its brown wings creating a cover over its abdomen. It is found all year round outside the caves.
- *Family Simullidae (black flies)* – these are from the Simulium species and surface streams wash in their larvae into the caves. These can be seen anchoring to the rocks in streams that are flowing quite fast, trying to acquire some debris while it's floating past. Among the flies in the Great Bedding are the adult blackflies.
- *Family Chironomidae* – this family is inclusive of the non-biting midges such as the Chiromomid blood worm larvae. Many of the flies found at Porth yr Ogof in the Great Bedding are Chironomid fly species.
- *Family Culicidae* – a common example of this is the Mosquito *Culex pipiens*. The majority of the ones found in the caves are females that have been inseminated and are overwintering in the caves. They have to exit the cave at some point in the spring to look for a blood meal, mainly from birds rather than humans. Each year provides many generations of these species.

• <u>*Sub-order Brachycera:*</u> These have bodies that are more roundly proportioned with antennae that are much shorter. Some families will be discussed below.

- *Family Heleomyzidae* – one of the common cave visitors is *Heleomyza captiosa* which is a Marsh fly. They normally enter caves during autumn and stay there until they are attacked by fungi or they starve. It is possible that they find dry places to settle in the dark zone during the winter and end up getting caught out with the reversal of air currents during the summer, with the corresponding rise in condensation and humidity. They may also be present in some caves all throughout the year, making it quite unclear why they are visiting the caves in the first place. The high humidity makes it easier for fungi to attack them. The younger adults can be found in the cave all through the year, mainly in areas that are shaded and close to the decomposing matter, however, the reason for this is still unclear. *H. capitosa* and *H. serrata* are quite identical to the eye, and their recordings are probably mixed up. The main source of nutrients for the larvae are sheep droppings and therefore are sometimes also referred to as dungflies (it is important to note that the true dung flies are members of the Scathophagidae family). Some fly species such as the adult Heleomyzidae flies feed on other flies from different species. Many of these flies have been recorded in the cave environment.

Figure 4.14: There are approximately 700 known Marsh fly species, where each has an association to a specific type of host and is found in habitats that are marshy. They commonly lay their eggs on the host animal from which the larva is going to feed on. After the maturation of the larva it is able to pupate within the shell of the snail that has been consumed.

Source: https://www.flickr.com/photos/rundstedt/4280574219

- *Family Phoridae (coffin or scuttle flies)* – these are referred to as flies however they mainly move by running around and are reluctant to fly. Their name is derived from their seeming reluctance to fly and their scuttling behavior, as well as their close association to dead animals and humans.

4.7.1 Order Lepidopterata (Moths and Butterflies)

Herald moth (*Scoliopteryx libatrix*) and the Tissue moth (*Trichosa dubitata*) are seen in high numbers in the entrances of caves. During this time in the cave, they also tend to move around a lot within the cave, such as Eglwys Faen. Both of these insects choose the cave as the best environment to overwinter, and in this period, they tend not to move around the cave. The Tissue moths tend to settle closer to the bottom on the walls and much closer to the cave entrance in comparison to Herald moths. Caterpillars normally feed on blackthorn and buckthorn.

The Herald moths normally stay under the ground for as many as 10 months, and the majority of the time it is found beaded in moisture in the dark zone. It may not be true that they spend the whole 10 months in the cave hibernating, as a section of this time could be spent developing their ovaries. Their heads are kept pointing away from the light. The caterpillars feed on poplar and willow.

4.7.2 Orders Trichoptera (Caddis fly), Plecoptera (Stonefly), and Ephemeroptera (Mayfly)

Insects in this order normally have two pairs of wings, however, they have a reduced size of their hind pair of wings, while in some of the insects the hind pair is nonexistent as in some mayfly. Caddies flies, stoneflies and mayflies have nymphs that inhabit streams, some have a number of generations each year, and there are other larger species where their nymphs live in rivers and some streams for a couple of years before they become adult flies. The adults do not have a long life span, as they live for only a few days. Percolating water washes their nymphs into caves, leading to the development of certain species within the caves, some going much deeper into the cave.

Mayfly for example Ecdyonorus and Baetis are commonly found in allogenic cave streams. The adults in this species make sure their wings are held upright while they are resting and they also possess three tails (sometimes two). All nymphs possess 3 tails.

Figure 4.15: These insects normally depend on habitats that are of good quality, in both terrestrial surroundings and directly in the water, so asto be able to finish their life cycle so as to establish populations that are healthy. The waterbody's health is dependent on many different aspects, the main ones being habitat and water quality.

Source: https://www.sciencemag.org/news/2018/01/muscular-mayfly-has-arms-airplane-wings

Stoneflies, for example,*Dinocras cephalotes* has been recorded in caves more frequently. Others include *Leuctra* sp. and *Nemouras cambrica*. The adults in these species are placed their wings on their back in a flat manner, while those in the Leutricidae family place their wings on their abdomen in a rolled manner. The larvae in this species have two tails that are quite long, and the adults look like the larvae but winged, however, some have lost their tails.

Caddis fly (P. geniculata and Plectrocnemia conspersa) larvae normally capture prey by spinning nets within the stream. They can make different types of nests all dependent on the condition of the water. This behavior of variable net spinning allows them to achieve the exploitation of many different microhabitats and prey. If there is wash in of a sufficient supply of food by the stream, the larvae and nymphs are able to continue surviving undergo, however, immediately they reach their stage of metamorphosis, and metamorphose into adults with wings, they have higher chances of dying because these adults mainly depend on plants as their source of nutrients. The adult yellow- spotted sedge (*Philipotamus montanus*) iscommonly found in caves. Adult *Stenophylax permistus* (large buff- colored caddisfly) normally enter the caves during the summer as a resting stage referred to as diapause.

4.7.3 Coleoptera (beetles)

There are actually many different beetle species that are cavernicoles. Many different species can be found in cave thresholds, however, the majority of them that are commonly found in caves are found underground, and these include members of the robe beetle (Staphylinidae) and ground beetles (Carabidae) families. In the United Kingdom, approximately 990 species have been recorded.

Choleva genus members are found in the thresholds of the cave feeding on fungi and detritus. During the summer, they form a cell made of clay to stay in, then reemerge in autumn. This is referred to as the summer diapause.

Trechoblemus micros is a small orange beetle that has blind larvae, and these have been located in Ogod Ffynnon Ddu, Porth yr Ogof and Otter Hole. The adults have eyes and they do not seem to have undergone evolution to survive in dark areas – despite the fact that they can only be found in caves. These are commonly found in British caves and are predatory.

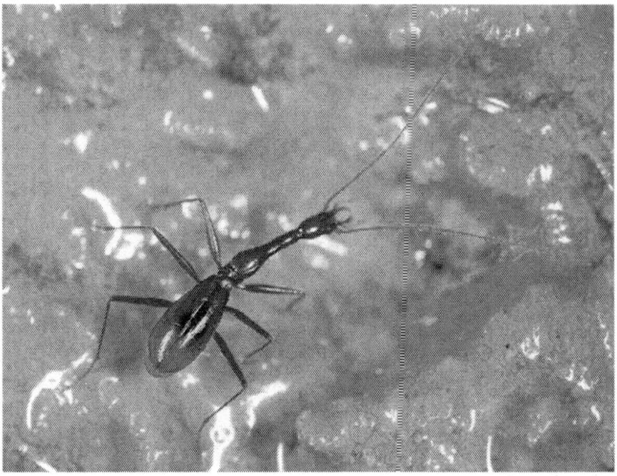

Figure 4.16: When beetles adapt to areas that are permanently dark such as caves, they tend to look like one another, having similar characteristics, such as loss of flight wings, pigmentation and eyes, spider-like appendages, great elongations, and compact bodies.

Source: https://www.esf.edu/top10/2018/10.htm

There are a number of rove beetle species found in Wales – *Lesteva pubescens* and *Quedius mesomelius* have been discovered in Porth yr Ogof. Diving and water beetles are also located in caves, and the majority of them end up there after being washed in during floods.

Other insects include fleas which are commonly found on bats. They leap from one bat to another even though they do not have wings.

- Spiders and Millipedes.
  - *Spiders:* The large *Meta menardi* is the most familiar cave spider. These spiders normally require space for them to able to spin their unique orb webs which are quite small in comparison to the size of the spiders. This spider is dark brown and quite large. They are troglophilic spiders that choose to live in darker areas or areas where there is a certain reflection of light. In the winter, they tend to live in the slightly deeper parts of the cave or places that have two close entrances near by. Some of them normally move towards the cave threshold during much wetter times of the year – the food web dynamics may change during the winter. These spiders also prefer vertical shafts, especially those that make iteasier for prey to fall in. The webs they create have a certain orientation away from the light, approximately 90° to the light. They build their webs in the shadow, for example in scallops and phreatic pockets. They do this because it makes it easier to trap the fungus gnats that are looking for shaded areas. Another reason for this is because their prey finds it hard to spot webs that are in shadows.

Figure 4.17: Majority of the adult spiders are highly sensitive to the light, making them photophobic, and that means they look for places without light to live in; mainly in tunnels and caves, however, there are times in which they are seen

outside mines and caves because they end up emerging from them around dusk to hunt. They normally hunt using one lasso line made of silk, which they use to swing down upon their prey.

Source: https://www.google.com/url?sa=i&source=web&cd=&ved=2ahUKEwjEjKym68TpAhVGVhoKHa_PCagQjhx6BAgBEAI&url=https%3A%2F%2Fen.wikipedia.org%2Fwiki%2FMeta_menardi&psig=AOvVaw09GCJClt-ygKt-FxlmMyD0&ust=1590146433049598

They never spin their webs across a passage unless they have no other surface they can use to anchor their webs to which are many a times parallel to the wall. Generally, spiders utilize their webs as supportive platforms, instead of actively using them for the trapping of prey, because the majority of their webs are not sticky. There are however some outlaying threads in their webs that act as trap wires that tell the spiders if there is prey close by on the ceilings or walls. They normally feed on flies, earthworms, slugs, millipedes and woodlice, and sometimes smaller *Metellina merianae* (however this evidence is not yet solid). These spiders have the ability to gorge themselves and they can keep their food stored in their webs by trussing it.

- Mites: These are smaller spider relatives. There many different mites that have been found living in Welsh caves, a number of the species being troglobites. One of the most common species is *Rhagidia spelaea*, and is mainly found in Ogof Clogwym and Ogof Ffynnon DduThe majority of the mite species are truly microscopic and they are hardly ever noticed. They normally live in many different environments – some prey on, for example, Collembola; some are parasitic on animals and plantsand some are free living. *E. loricatus* and *Eugamasus magnus* have been located in Eglwys Faen and Porth ys Ogod. Of the mites found in Welsh caves, there should be some water mites that can be added to the list. The Rhagidia species in Ogod Ffyonnon Ddu that normally feed on Collembola are found in many areas but they are hard to find, just like many other cave creatures.
- *Millipedes and other Myriapods:* The small feet (Pauropoda), are approximately 1mm long and they are eyeless creatures that have a certain similarity to centipedes shape-wise, having approximately 10 to 11 pairs of legs. They are mainly found in the soil receiving their nutrients from decaying plant matter.They have a certain similarity to springtails, however, when you disturb them, they end up running backward in a fast manner. Researchers have

thought of them to be scavengers. Records from caves all over the world show that they are found in detritus accumulations in the cave thresholds. There are also the Symphyla species that are much larger however still under 1.2cm. These are centipede-like organisms that are white and blind, and they also feed on decaying plant matter that they find in the soil. The younger adults have 6 pairs of legs, however, after a number of years, at each molt, they add an additional pair of legs, and in the end, the full -grown adult has approximately 12 pairs of legs. There have been records of *Scutigerella causeyae* from Ogof y Ci. These species have been found in Guzzle Hole, mainly in the dark zones of the caves.

Figure 4.18: Above is a photo of the *Causeyella youngsteadtorum*, and this is one of the common cave millipede species in the United States. It is ghost white in color, and it was first discovered in 1977 by Jean Youngsteadt and Norman Youngsteadt. It was however disregarded as a new species up to the year 2003. It resides in approximately 7 different caves in Searcy and Boone countries in Arkansas.

Source: https://bugguide.net/node/view/630255

Diplopoda (twin feet) include millipedes, and these have 2 legs for each segment. Millipedes have been in existence for more than 430 million years, and this makes them the oldest terrestrial animals. Researchers have been able to prove that the ones we see today are highly similar to the ones that were there in the past. They have undergone adaptation allowing them to push their way through leaf and soil litter when they are feeding on decaying plant matter such as wood. These creatures are quite docile and they tend to curl up at any one point if they are threatened or feel threatened, however, they also have passive defense mechanisms, including noxious and caustic chemicals. There are some that release cyanide when they are threatened, and this normally smells like almonds that have gone bitter.

When their eggs hatch, they release the first setthat have 6 legs, and these can be confused for insects. These undergo molting several times, being able to gain legs and segments every single time they molt until they are adults. Certain species normally die after they mate and therefore end up living for a maximum of one year, and there are also some other species that live for as long as 5 years. The fact that some species have short life cycles means that they find it hard to find adults to mate with in certain times of the year. They have normally much thinner shells in cave forms, in addition to having more sensory bristles that those that are seen on the surface.

5

Grassland Insects

CONTENTS

5.1 INTRODUCTION

A number of invertebrates are found in grasslands whether they are above or below the ground. They collectively include a number of insects and are represented in different trophic levels. Many of the taxa include insects and they scarcely intrude interests in relation to conservation. Their abundance and biomass and their importance is acknowledged in ecological functioning. Insects are major components of the fauna in grasslands and play a role that is substantial in processes in the ecosystem such as the cycling of nutrients and also the pollination process but their role is often overlooked. They are dependent on the diversity of plants and are the most abundant in terrestrial ecosystems including grasslands. They contribute to shaping grasslands through processes that are top-down ,for example, they can modify richness of plant species by altering competitive dynamics between plant species and likewise the composition of the plant community will also play a bottom-up role in structuring communities of insects and these alsapply to abiotic factors such as temperature and the availability of water. Plants found in grasslands can also be impacted by changes in climate. Changes in rainfall can also affect plant and insect communities as a whole and it impacts on their interactions since it is climate- sensitive (Hiltpold & Rostás, 2017).

Grass covers a large portion of the surface of the world than any other type of vegetation. It has a simple architecture and the flowers lack a color that is conspicuous colorful structures and therefore receive very little attention. They are still of economic importance that is relevant. They are composed of cereals thatare consumed by a large population in the world and they also contain the world's worst weeds. Insects in general and also those found in grasslands provide a diversity that is often overlooked due to some of them being inconspicuous. Most of the studies that have been done on grasslands revolve around insect pests and the impact they have on agricultural crops. Herbivores that affect rangeland productivity or ectophagous insects that are associated with grasses. Few studies have been done on insects that are endophagus thatfeed inside the shoots of grasses although these species are mostly expected to be specific to certain grass habitats. Monocots and dicots are present in grasslands and they are referred to in most studies and therefore interactions between plants and insects have been studied in vegetation contexts that are broad, that is grassy unless invertebrates are the Centre of focus and interactions in rangelands (Hiltpold & Rostás, 2017).

5.2 SPECIAL FEATURES OF GRASSES AND GRASSLANDS

5.2.1 Grass

Grasses differ from dicotyledons whose structure is simple, has complex branching that is rare; they also lack thickenings that is secondary and intercalary meristems thatact as growth substitutes for buds that are at the end and are protected with hard sheaths of leaf. In addition, most grasses lack a variety of compounds that are secondary toprevent herbivoes from attacking dicotyledons. If such chemicalsare present, they exist in low concentrations. Alkaloids are present in less than 0.2% of the species of grass. Cyanogenesis is very restricted; many chemical compounds classes characteristic of other plant families are completely absent.

Plants that are medicinal and toxic are also rare among the grass species and there is little information known about the negative effects of a few toxins on herbivores. Some insects reject plants with high concentrations of secondary compounds and therefore are restricted to certain grasses. Deterrents also govern the choice between grasses and shrubs which are to be fed on by insects. Food preferences within grasses also appear to be mostly due to nutritional factors of the host plant. High silicate content of grasses reduces the palatability of grasses. Silicate and lignin have functions that are or maybe nondefensive. Xerophytic plants are thickened by compounds such as lignin. They prevent processes such as wilting and provide structural support to plants. Leaves thatare soft and nutrient- rich from grass seedlings have secondary compounds with higher concentrations of compounds that are secondary and have biologically active than older leaves do which are protected better by physical defenses. Seedling compounds prevent feeding by graminivores that are oligophagous and these include acridids and caterpillars but other species ,for example, cereal leaf beetles prefer shoots that are young for oviposition and feeding. Compared to other herbivore families, the percentage of proteins in grass is less. Age is a major characteristic also affecting the availability of nutrients in grasses. Internodes that have been developed have nutrient concentrations that are elevated such as water, proteins and minerals and reduced concentrations of the factors that are responsible for plant resistance effects such as raw fiberand silicate, so nutrient availability decreases significantly from the top to the root base as the age of the internode increases (Allaby, 2009),

5.2.2 Grasslands

Grasslands ecosystems cover up to ten percent of the surface of the earth and occur across a wide range of changes in climate. Grasslands are found where there are fifteen to seventy-five centimeters of rainfall annually and therefore do not support forests but is above that which occurs in deserts. Grasslands are mostly found in areas with temperate climates. Some of the most common grassland areas around the world include the steppes in Europe and Asia, the prairies in the Midwest of the United States, pampas in South America, veldt in South Africa, savanna in most countries in Africa and the downs of Australia. In a country like India, it is found high in the Himalayas. The rest of India's grasslands are mainly composed of the savanna and steppes. Formations in steppes mainly occupy areas that largely have sandy and saline soils.

Grasslands are regarded as one of the terrestrial biomes that are the most threatened. Grasslands are important in maintaining the landscape level biodiversity and they support species that are rare especially those with populations that are fragmented and small and therefore deserve conservation. Insects in grasslands arewidely appreciated together with their roles in sustaining ecological function diversity. Insect diversity is influenced by many factors and mechanisms that are either direct or indirect.

Grasslands generally grow in areas where there is less moisture and they also contain plants that do not grow much since they do not receive a lot of rainfall. They can survive in such areas because they have deep roots within the soil and therefore they receive moisture from the soil. The type of vegetation in grassland ecosystems is dependent upon the amount of rainfall the area receives. Different types of grasslands will differ in terms of their animal species similarly the species have also adapted to the climates they live in (Allaby, 2009).

Grassland ecosystems can also be divided into five groups temperate, tropical, flooded, montane and desert grasslands. Tropical grasslands receive 50 to 130 cm of rain. They experience both days that are rainy and dry. As mentioned above they are referred to as the savanna and therefore they contain shrubs and small trees that are dry in nature. They contain several animal species including insects

Temperate grasslands receive rainfall of about 25 to 75 cm. These type of grasslands extreme face climates. During cold seasons, the temperatures can reach grasslands that are flooded to 0 degrees Fahrenheit. During the summer temperatures, it can reach up to 90 degrees Celsius. Precipitation in these grasslands is mostly in the form of dew and snow. For instance, some vegetation that grows here includescacti, sagebrush, perennial grasses, buffalo grass clovers and indigos that are wild.

Flooded grasslands have water all year round. These grasslands contain numerous vegetation that grows in water. Various water birds migrate to these areas while some are residents of it. The Everglades is the world's largest flooded grasslands. These grasslands also feature a variety of animal species. Plants found here have structures that arerosette, abundant pilosity and structures that are waxy for instance, the northern Andes contain this type of a habitat.

Desert grasslands are the type of grasslands that separate the true desert of lowlands and montane grasslands. Such grasslands receive low precipitation and they result in grasslands that are hot and dry. These grasslands are scattered mostly as they are dependent on the areas that receive rainfall (Allaby, 2009).

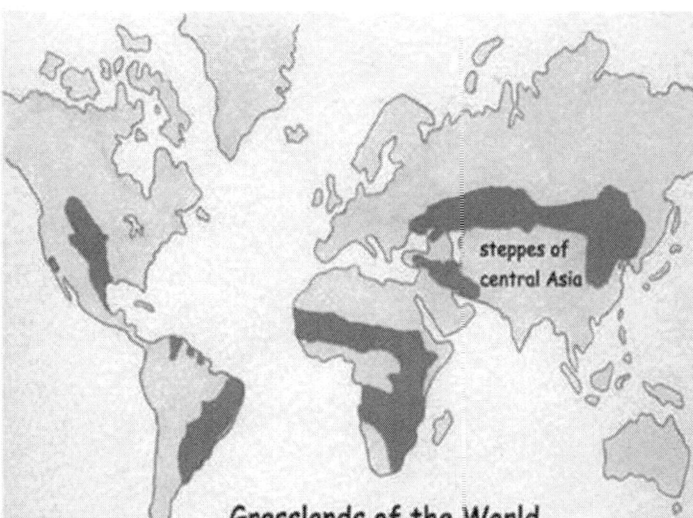

Figure 5.1: Coverage of the steppes grassland globally.

Source: https://www.exploringnature.org/db/view/1716

Figure 5.2: Acacia savanna in Taita Hills Wildlife Sanctuary, Kenya.

Source: https://en.wikipedia.org/wiki/Savanna

Figure 5.3: The pampas at eye level in Brazil.

Source: https://en.wikipedia.org/wiki/Pampas

There are majorly two types of grasslands and these are either tropical or temperate. Temperate grasslands include the Eurasian steppes, the prairies of North America and the pampas of Argentina. Tropical grasslands include hot savannas which are found in sub-Saharan Africa and the North of Australia. Grass species include Rhode grass and red oat grass found in the tropical savanna. Temperate areas have needle grass and galleta. During monsoons, the grasslands become coated with flowers that are wild such as the hyssop, yarrow and the milkweed. Grassland plants have adaptations that allow them to survive in droughts, fires and grazing in the habitats they share.

Figure 5.4: The read oat grass *Themeda thiandra.*

Source: https://www.feedipedia.org/node/367

Figure 5.5: The Rhodes grass *Chloris gayana* which is commonly found in tropical and subtropical regions.

Source: https://www.feedipedia.org/node/367

Figure 5.6: The needle grass or stipa which is commonly found in the savanna and certain species are found in the prairies.

Source: https://en.wikipedia.org/wiki/Stipa

Figure 5.7: The galleta grass.

Source: https://greatbasinseeds.com/wp-content/uploads/2013/09/galleta-grass.png

Grassland ecosystems are mainly inhabited by grass, plants that are not woody. It is also a landscape that is transnational as they are dominated by grass or no trees in the areas where there is not enough for a forest and also not too much for a forest. There are different components ecosystems inhabited by grass and they include abiotic componentsthat are non-living such as carbon, hydrogen, Sulphur, nitrogen and phosphorus. It also includes components that are biotic which are living.

1. Producers- they primarily produce food and are grasses such as Aristida, Cynodon, Digitaria, Desmodium, Setaria and others. Herbs and shrubs contribute to food production primarily in grasslands where they exist.
2. Consumers- in grasslands consumers are at three levels. The first are primary consumers and they feed on directly on grass and this group includes animals such as cows, buffaloes, goats, rabbits and mice; these group similarly includes insects, termites, centipedes and millipedes. The second level is for secondary consumers and they include animals such as snakes, lizards, jackals, foxes among others and they feed on consumers that are primary as mentioned above. The last level is tertiary consumers such as hawks, eagles and vultures found in a grassland ecosystem and they feed on both primary and secondary consumers
3. Decomposers are the last group of biotic components. Grasslands have organic matter which is decomposed by microorganisms like actinomycetes; fungi such as Mucor, Aspergillus, Rhizopus, Penicillium and Cladosporium; anaerobic and aerobic soil decay. They increase the soil fertility by improving the soil profile and also adding minerals to the soil.

Figure 5.8: An image showing the different levels of interactions between biotic components found in a grassland ecosystem

Source: https://scioly.org/wiki/index.php/Ecology

Grassland ecosystems have different functions. Their primary function is productivity. The producers fix energy from the sun and produce organic matter that helps with soil fertility. It also functions to provide forage for livestock. It also functions to protect and conserve the resources of soil and water. It also functions as a habitat for wildlife, other flora and fauna. All these generally increase the aesthetic value of such areas. The functional aspect of grasslands can be studied in two ways (Hiltpold & Rostás, 2017).

The first is an ecosystem food chain. An important feature of ecosystems is that organisms from one level are consumed as food by organisms from another level. A series of these levels form a food chain. The food chain does not follow a chain that is linear in a grassland since an organism may feed on multiple organisms from the same food chain or it may also feed on organisms from food chains that are different thus a food chain that is interconnected is a food web (Fleisher, 2013).

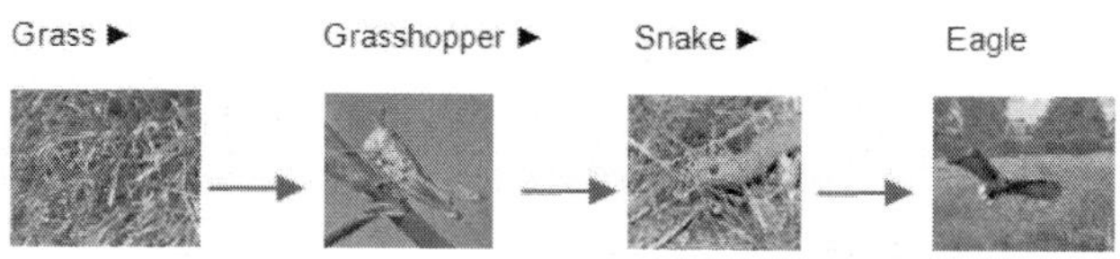

Figure 5.9: An example of a food chain in a grassland ecosystem.

Source: myriverside.sd43.bc.ca/morgank-2014/biology-project-the-australian-tropical-grasslands/sample-page/

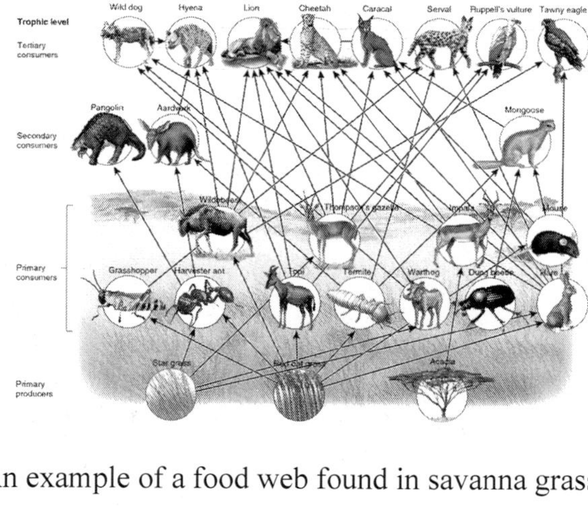

Figure 5.10: An example of a food web found in savanna grasslands.

Source.https://tropicalgrasslandsavannahksmith.weebly.com/food-chain--food-web.html

The second is nutrient cycling in an ecosystem. For the success of any ecosystem, it is important that the constituent materials move in a manner that is cyclic. Producers take up soil nutrients and minerals elements from the soil and afterward convert them into a form that is organic and after these minerals pass through the different trophic levels they again return to the water or soil. Grasslands also have economic importance. First, they are important I the maintenance crops of many herbivores that are either domesticated or wild such as horses, mules, cows, pigs, sheep, buffaloes, camels, zebras among others. These provide milk, other products and some of them can be used as a means of transportation by human beings. Grassland ecosystems are composed of a mixture of grass, leguminous species, herbs and shrubs and therefore it can contribute to its preservation and conservation. In temperate grasslands, herbivoresaredominated by insects such as poikilotherms or heterotherms but large ungulate herbivores for example homeotherms dominate grasslands that are tropical. Herbivory by homeotherms results in comparatively higher rates of consumption but secondary production that is lower. Insects such as caterpillars and acridids ingest their own body weight or more food daily, so that the ratio of biomass removed to grazer biomass is higher. Lack of differentiation between outbreak and non-outbreak systems or between ecosystems with herbivory that greatly fluctuates and is chronic making estimates off herbivory difficult since damage made by grasshoppers and grass- feeding stem borers or outbreaks of forest pests can fundamentally change the relative importance of detrital and grazing lands (Fleisher, 2013).

5.3 INSECT COMMUNITY STRUCTURES IN GRASSLANDS

5.3.1 Ectophages and Endophages

Mortality of the critical mature stages of insects is affected by the niche of insects that are phytophagus. Ectophages feed on leaf tissues which are external by chewing, scraping or sucking are different from endophytic feeders which include leafminers, gallers and borers. The feeding categories can affect the extent to which refugees can be occupied by herbivores protecting them from attack by parasites and importance which is relative of mortality which is specific by stage. Chewing of grass foliage by insects primarily belong to Orthoptera, Lepidoptera, Coleopteran, Hymenoptera and Phasmida. Specification appears to be more in Orthoptera than in other groups. Insects that feed on sap from grasses are Homoptera, Heteroptera and Thysanoptera. Herbivores that bore on stems primarily belong to Diptera and mainly Cecidomyiidae and Chloropidae; Hymenoptera such as Cephida and Eurytomida; Homopteramainly Pseudococcida; Lepidoptera mainly Pyralida and Noctuida; Coleoptera mainly Cerambycida and Mordellidae and mites (Collins, 2012).

Comparisons of the association between plants and insects that feed on grass support the hypothesis that endophage to ectophage ratios are high particularly on plants that are heavily defended because endophages such as gall formers are better able to selectively feed and can manipulate the development of tissues to avoid defenses that are either chemical or physical on the other hand, ectophages are mainly feeders that are unselective. The specificity of the feeding in endophages results in separation that is spatial and temporal in relation to parts of the host plants that are nutritious and consequently these insects are greatly affected by variability in host plants or change in the nutrition value of parts of plants.

The paradox can be solved by differentiating between effects that ecological and evolutionary. Specificity in terms of ecology is a prerequisite that is ideal for rapid speciation and price showed that the chemical diversity of potential hosts apparently forced specializations to parasites resulting in a greater diversity of species.

In plants, diverse defenses greatly reduce the occurrence of attack by each of the gallers that are specialized, with time these complex defenses are likely to be prevailed over by a number of species that are specialized and also evolve in steps. The arguments can be useful in the prediction of the ratio of endophages to ectophages that is low in grasses compared to dicots since dicots are characterized by compound diversity with a chemical nature. The limited available data is also consistent with the hypothesis in large insect fauna volumes for example the oaks or bracken which are made up of species of gall- forming and leaf -mining. On the other hand, communities of herbivores in grass seem to be controlled by ectophages (Collins, 2012).

The perennial samples of *Dactylis glomerata* consisted of three times as many ectophagus species as endophagus species. The ectophages group consists of insects that chew foliage and do not feed on a single species, on the other hand, those that feed on foliage on other host plant families are more specialized. Adaptions of grasshoppers that are oligophagus to the defenses of one species of grass appear to be sufficient allowing them to feed on others as well. In plant hoppers that are sap feeders. which primarily develop on grasses, monophagy is widespread and in leafhoppers that are in grasslands, oligophagy is common. Selection and host finding of insects in acridids is mainly dependent on vision, chemoreception contact, and identification of physical characteristics such as trichomes of leaves or simply an assessment based on biting. The causal basis of host grass is known for the endophagous shoot fly,*Oscinella frit* which distinguishes between the epicuticular wax components of host and non-host species of grass. In general, the host plant preference of oligophagus grass differs significantly between species of grass.

5.3.2 Insects in Grasslands

There are abundant species in grasslands. They include bumblebees, crickets, butterflies, locusts, grasshoppers and beetles. In North America, the prairies mole cricket is one of the most common species and can grow up to the length of two inches. They are uncommon, nocturnal and they spend most of their lifetime burrowing underground in tallgrass prairies. Grasslands are regarded as one of the terrestrial biomes that are the most threatened. Grasslands are important in maintaining the landscape level biodiversity and they support species that are rare especially those with populations that are fragmented and small and therefore deserve conservation. Insect diversity in grasslands is widely appreciated in addition to their role in sustaining their function within the ecosystem.

5.3.2.1 Key Insect Groups in Grasslands

Orthoptera: They order includes grasshoppers, locusts, crickets and related insects such as wet as and katydids. Ortho means straight and ptera means wings. The first pair of wings are elongated and leathery. They also have filiform antennae. They have well -developed cerci. In females, the Orthoptera have long ovipositors. Species of orthoptera sometimes play the role of pests in grasslands as they can wipe large fields of grass and even agricultural grasslands since they can move in large swarms. They can also create biofuel when they break down lignin and cellulose without producing greenhouse gases. The sword- bearing conehead grasshopper is common in the prairies in both Canada and the United States. Their nymphs and adults feed on the leaves and other parts of plants. Their common predators are birds. They are insects with the incomplete metamorphosis. Other than the prairies, they are found in agricultural grasslands, pastures, meadows, fencerows and along railroad vegetation. The handsome grasshopper lives in grasslands, wet meadows and prairie remnants. They live for one season and lay their eggs in the soil. It goes through five growth stages and sheds its exoskeleton at each stage. Their nymphs and adults both feed on the bluestem grass and other grasses. Their predators include different bird species.

Figure 5.11: The roesel bush cricket.

Source: https://www.naturespot.org.uk/species/roesels-bush-cricket

Hemiptera: They are true bugs. They include groups of insects such acicadas, aphids, planthoppers, leafhoppers and shield bugs. Their front wings are thickened, they have a leathery base and a tip that is membranous as they rest their wings are flat over the body. They also have mouthparts for sucking and piercing and beak that arises from the front of the head. Many of the insects have a triangular shape referred to as the scutellum. They feed on plants and they use their mouthparts to feed on sap from plants. Some are hematophogus while others feed on other insects. Some of the grasslands are inhabited by aphids which can also be agricultural pests.

Most of the group of insects are herbivores. Some of them are phytophagus such as the cicada, leafhoppers, treehoppers, plant hoppers, aphids, whiteflies, scale insects and others. Others can be monophagous and can attack a specific plant host and others are oligophagus and feed on a new species. They sometimes also change the plant mix and they feed on plant roots or seeds. Depending on the time of the year, for example, aphids spend the winter on the plants hosting as eggs. They spent time on plants that are woody as pathogenically reproducing females.

Figure 5.12: A twig wilting bug (Coreidae) piercing and sucking sap from a Zinnia.

Source: https://alchetron.com/Coreidae

Coleoptera: Coleoptera are a family of beetles. The front wings are modified into hardened or leathery elytra meeting a straight line down the back and covering membranous hind wings. It has mouthparts for chewing and is a very large and diverse order of insects. Beetles exist in a variety of habitats including coastal and freshwater. They exist where there is vegetative foliage from trees, flower barks, leaves and underground near roots. In grasslands, they play different roles such as parasites, some as pollinators and others as pests and finally, they play a role as biodiversity indicators to measure regional patterns of biodiversity.

Figure 5.13: An Israeli Copper Flower-Chafer (*Protaetia cuprea ignicollis*).

Source: https://www.biolib.cz/en/image/id262734/

Some of the insects found in the savanna grasslands include the bushman arrow poison beetle and the rhus flea beetles. The bushman beetle has distinctive small black dots. It uses a host plant to lay its eggs. The larvae which is parasitic can feed on the host tissues. In grasslands, the bushman beetle contains a protein toxin that is used by Bushmen to coat their arro heads when they go hunting. The rhus flea beetles on poisonous plants that intern make them poisonous and therefore making them useful in hunting. Other beetles found in grassland biomes include dung beetles and they break dung piles which they encounter. Their larvae exclusively feed on dung and their adult form may also feed on mushroom therefore decomposing plant matter. They are useful in savanna grasslands as they role dung preventing them from hardening hence impeding plant growth. In the prairies there exists the goldenrod soldier beetle and it related closely to the lightning bug. Such bugs feed on nectar, pollen and insects that are small. The dogbane leaf beetle is also common in the prairies. They lay their eggs on the ground and on the milkweed plant. Their larvae move through the soil and consume roots and feed on them. They avoid some of their predators by producing a smell that is foul. It is present in the Atlantic coast to Texas and the north of Saskatchewan in Canada. It is also common in the Illinois prairies. Other beetles in the prairies include the milkweed longhorn beetle. They are also found in meadows and other grasslands. Their female lays their egg on or near the milkweed plant. Their predators include birds. Their larvae bore into roots or stems of milkweed and their adults feed on buds and leaves. They are found throughout the United States.

Lepidoptera: It includes insect species such as butterflies skippers and moths. In this family of insects, the wings are covered with scales and they come off as dust. They have sucking mouthparts coiled proboscis. They also have large compound eyes. Butterflies and moths play the role of pollination and as food in the food chain and their larvae is also considered as a problem to vegetation. They are species that are found in all continents and they inhabit ecosystems that are terrestrial and it ranges from deserts to rainforests and also lowland grasslands. They exhibit lifestyles that are either symbiotic, phoretic or parasitic. Adult butterflies and moths are preyed upon by other insects such as dragonflies, birds, lizards and spiders. Insect -eating birds are their predators and they have developed a number of defense strategies. Some species such as the monarch butterflies are poisonous predators. They can also act as pests. The monarch butterflies are found in the prairies and lay their eggs on milkweed. They are long-distance travelers and travel with the North American monarchs and range from the south of Canada to the north of America. The tiger swallowtail butterfly is also common in the prairies. The Virginia ctenucha moth feeds commonly on the grass and is found within the prairies. Others include the star borer moth is found in the prairies of the Chicago region and some parts of Illinois. Their feed on the inner stem and the root base. Their active life is affected by temperature, humidity, precipitation, wind and moon phases. Their predators include rodents, skunks, woodpeckers and flies and wasps that are parasitic. The Leonard skipper butterfly is found from Canada to

Alabama and also the west as far as Minnesota and Missouri. It lays its eggs in grassy areas. It resembles a moth and has scent scales on its wings. It flies in August and September. Their larvae feed on the leaves of grass, they hibernate. The adults fly from august through October. The larvae eat switchgrass, love and bentgrass. The adults feed on the plant nectar and they include thistles, asters, blazing star and teasels. The northern mole cricket lives in the eastern and the central of the United States and also the west to the Great Plains and the south of Texas. They are hairy and brown and they live in grasslands, meadows and prairies areas where there is the growth of grass. The adults and nymphs feed on seedlings and grass. Their predators include birds, ground beetles, assassin bugs, mice and tiger beetles.

Figure 5.14: The monarch butterfly.

Source:https://www.needpix.com/search/viceroy%20butterfly%20on%20grass

Hymenoptera: Hymeno means the god of marriage. They include insects such as ants, bees and wasps. The insects have four wings that are membranous, the hind wings are smaller and have a row of hooks for joining the front wings. The mouthparts are modified for chewing and lapping. It has a developed ovipositor and it is sometimes modified into a stinger. They are hard bodied active insects with compound eyes that are well developed and they are often have thread waists. Another common member found in plains, woodland areas and grassland habitats is the clymene moth and it feeds on a variety of plants.

Figure 5.15: A digger wasp, Sphex pensylvanicus.

Source: https://pbase.com/crocodile/image/34100054

Termites thrive in savanna grasslands. They are hemimetabolous this means they don't enter a larval stage in their development: they go from eggs straight to nymphs and gradually develop into adults. They live in large colonies, some of them defend the colony, others are workers and they ,therefore, build and maintain the colony. On surfaces in the savanna, their colonies appear as grassy mounds and have an influence that is positive on local animal and plant activity. Ants live in vast colonies either underground or on the ground which each playing a role in their colonies. The rusty patched bumblebee is an endangered species found in prairies and other grasslands. They collect nectar and transfer pollen.

Diptera: They are commonly referred to as true flies .they have one pair of membranous wings that are functional. They also have large compound eyes and mouthparts for sucking, piercing or sponging. They have antennas rising between the eyes and they are often short and segmented three times. They play an important role at various trophic levels both as consumers and prey. Their larvae can be herbivores, scavengers, decomposers, predators or parasites.

Figure 5.16: Anthomydiie flies.

Source: https://www.wikiwand.com/en/Anthomyiidae

Odonata: The order Odonata is made up of dragonflies and damselflies. Suborder Anisoptera is composed of dragonflies. It possesses hind wings that are broader at the base and wings nearly horizontal at the rest. The Aeshnidae include darners while the Libellulidae include skimmers. The sub-order Zygoptera is composed of damselflies. For this group, the front and hind wings are similar in shapes and the winds are held together above the body. Under this subgroup, there are Calopterygidae which are broad -winged damselflies and Coenagrionidae are damselflies with narrow wings. They are commonly used as bioindicators in water ecosystems.

Figure 5.17: Damselflies in a copulatory wheel.

Source: https://en.wikipedia.org/wiki/Odonata

Some of the most common types of low land grassland dragonflies include the common archtail, the scarlet percher, and yellow striped hunter. Damselflies include the common blue tail, the aurora blue tail. In the prairies, some of the common insects from this group include the black- winged damselfly and the ten spot dragonfly. The ten spot dragonfly is a predator that is feared. Their larvae are known to kill and eat small fish while adults can also kill hummingbirds. They also kill mosquitoes in their larval stage and also their adult stage.

5.3.3 Ecological Functions of Insects in Grassland Ecosystems

Insects represent a component of biodiversity that dominate most terrestrial ecosystems yet their role has largely been neglected in most studies that are being done especially the role of their biodiversity in the cycling of nutrients and the functioning of the ecosystem in general. In terrestrial ecosystems insects for examples those in grasslands function as herbivores, pollinators, seed disperses, parasites, predators, detritivores or ecosystem engineers. Despite all their roles, their roles in nutrient recycling is not recognized universally and largely due to their biomass in comparison to other animals and plants. Insects affect nutrient cycling both directly and indirectly. Direct effects include the reduction of NPP by herbivores and break down of litter by detritivores. Other direct effects include engineering of the ecosystems. Indirectly insects affect plant or insect populations in their plant interactions through herbivory. Interactions among insects and other organisms also affect the functioning of ecosystems. The main consequences of interactions within grassland ecosystems include herbivory, phytophagy and mutualism including pollination, seed dispersal and the protection of plants. For herbivory, the direct effects on the storage of carbon and cycling of elements including NPP is well documented. Insect herbivore consuming material that is living and therefore will have a direct effect in NPP. Herbivory on roots, shoots, flowers and seeds will lead to changes in the storage of carbon through honeydew production, production of methane in the guts of insects, respiration, defecation, inducing the fall of leaves, reducing crops that are standing and increasing nutrient leaching from foliage. They affect the composition of plant species through a change in succession that is direct, affecting the diversity of plant species and it also affects rank abundance relationships. Nutrient cycling is also affected through mechanisms such as the alteration of food webs and the resource allocation of plants. Mutualism also affects nutrient cycling, for example, pollination and seed dispersal directly change carbon storage and the cycle of elements. Protective mutualisms directly change the carbon storage and the cycle of elements (Allaby, 2009).

Insect roles, for example, those that are detritivores also directly change the storage of carbon and cycles of elements by reducing crops that are standing, removal of dung and carrion and from the breakdown of litter and wood. Ecosystem engineering by insects directly changes the storage of carbon by moving nutrients. They also modify the conditions of habitats for other organisms by mixing substrates and microorganisms, reducing particle abrasion, modification of abiotic conditions, plant growth increments and water infiltration. Predators alter interactions within a food web through trophic cascades and the effects of the food web in soil. Microbial feeders change carbon storage directly for example by changing the rates of decomposition. They also alter interactions of food webs when they graze on fungal hyphae, affecting plant -mycorhiza interactions and interactions between organisms in the soil. Agents of dispersal and in this case insects disperse other organisms and examples include the distribution of the spores of Arbuscular mycorrhiza.

Hypothesis on nutrient cycling states that insects that are herbivorous increase the cycling of nutrients returning nutrients fixed in plants back to producers at a rate that is fast than the nutrients would flow otherwise through processes of senescence, the fall of litter and decomposition. Another hypothesis also states that low levels of herbivory actually increase the productivity of plants and the intensity of herbivory is shaped like a hump. In mutual interactions, pollination and the dispersal of seeds are services insects provide for plants. There are other protective mutualisms for example those that exist between ant and plants. Plants employ insects to defend them against other herbivores for example by providing them with domatia or by attracting them with emissions that are volatile. Pollinators and dispersers affect the composition of plants especially in their diversity (Collins, 2012).

5.3.4 Insect Numbers Per Grass Species

In an analysis done on plant and leafhoppers occurring per grass species, both ectophages: the leaf and planthoppers and endophages such as gall midges, phytophagus chalcids and cephids had five to twelve times more types on grass that are many and large than on grasses that are small and rare. Insect attacks on annuals were observed only seven times. 93 leaf miner species were known to attack at least one grass species. Records on plants attacked by insects lacked in only 14 out of the 45 species as observed. Grass with long shoots that was attacked with only one species were the phragmites with twenty-two leaf miners with phalaris as fifteen and genera with shoots that are medium -sized with numerous species. A study done by H. J Greiler and Tscharntke in Germany made this confirmation by comparing the endophagous stem borer community in 15 species of grass. The number of insect species on 10 perennials increased with the mean of the shoot length but five annuals were not attacked by the gall midge. According to the two variables, the shoot length and the life cycle dichotomy have the possibility of being used to predict the richness of species and ninety-seven percent of the variance. On average thirty-four insect species were found in every perennial grass species. The phytophagous species were the main species attacking small grasses. The taller grass species were attacked by chloropid flies, cephids, moths, beetles and small flies (Allaby, 2009).

The common reed *Phragmites australis* has the longest shoot by far and is attacked by about a hundred species of insects of which 93% are monophagous. Flush crash cycles of moths that are stem boring which damages red shoots influence interactions of food webs and the abundance of the whole insect community that occupies the P.*australis*therefore increasing insect diversity and the mosaic of habitats. Insect distribution among grass species can be explained in the following way, the length of the shoots is correlated presumably with food supply heterogeneity for example for steam borers that are large on can only feed on plants that havthick stems. Furthermore, long shoots should not only be a predictable source but also a resource that is large enough to allow specialization of parts of plants. Simple plants with simplicity in their structure are rarely attacked (Allaby, 2009).

5.3.5 Biotypes of Grass and Insects

Variation of genetics within species of grass can be great in not only the variety of annuals with inbreeding that is related but also in other species. Variability in palatability between perennial ryegrass strains is greater in comparison to the variability among many pasture species; a test of 45 varieties of ryegrass that is perennially screened for susceptibility to attack by the frit fly showed a number of responses. The dipterous larvae that were stem boring which attacked the shoots of a variety of Italian ryegrass from 13 species were negatively correlated with the content of silica which presumably influences female's choice of oviposition site. In a different example, insects attacked the German variety of thcocksfoot significantly more often than they attacked the variety from Poland. Only 5 % of the *D. glomerata* shoots grown by commercial seeds were attacked by ten species that are endophagous but similar insects attacked twenty to fifty percent of the shoots belong to wild biotypes. Grass developed from a mixture of seeds for grass from grasslands that have developed naturally with respect to the programs for nature conservation. This is in two ways ,for instance, fields that are sown have reduced diversity in terms of the plants that are available and this leads to a reduction in the diversity of insects. The second is the origin of commercial seed genotypes is different from those that are naturally established in a region. Such seeds may also be bred for insect attack resistance. They also don't reflect the different genotypes that occur naturally within and among populations (Hiltpold & Rostás, 2017).

5.3.6 Interspecific Competition and Facilities

Elevation of the population of grasshoppers in an experimental study in tall grass in the prairies did not have competition that was interspecific observed for example for food resources. Fluctuations in the density of grasshoppers did not affect other grasshoppers or species of Hemiptera and Coleoptera. Competition only happened in only two species of herbivore mired bugs of the fifteen that were examined in a study. There is little evidence that supports competitions that is interspecific by avoiding plants that are infested heavily. A five insect larvae community that all bore into *Spartina alterniflora* stems of grass, hence using similar resources have been analyzed using experiments. Insect concurrence was much rarer than was expected because their larvae sometimes fight to death while within the stem. The lepidopterous larvae of *Elachista* species lead to dying of delphacid eggs by movement through grass stems or leaf sheaths of *Dactylisglomerata* or *Holcus mollis*. The *D.glomerata*stems tunneled by cephids or beetles which bore through several nodes are attacked much less often by other insects of the guild from stem borers according to published data by Wesserling and Tscharntke.

In the event there are mass attacks, moths that are stem boring on *Phragmites australis* may lead to the destruction of many shoots, therefore, getting rid of food resources of the gall makers infesting the top of the shoot leading to a reduction of their numbers. When ten insect herbivores were analyzed, that are stem boring on *Calamagrostis epigeios,* the interspecific competition was a factor not found to be specific and this was according to Dubbert and Tscharntke.

Herbivore facilitation contrasts with a competition that is interspecific and both effects can be solved by stimulating changes in the growth of plants. Most of the time feeding decreases the quality of food in the host plant such as when leaves regrow from dicots not only restores the assimilation capacity and competition but also increases its protection in relation to herbivores. In the grasslands of Serengeti however, wildebeests that convert senescent grasslands into a community that is highly productive and therefore prepare grasslands for exploitation that is increased by other vertebrate herbivores. Similarly, the damage is caused while attacking the *P. australis*by the stem- boring moth converting shoots that are resistant into those which susceptible, thereby facilitating gall midge population growth.

5.3.7 Interactions between below Ground and above Ground Herbivory

Across the grasslands of North America the mean root to shoot ratio varies in the ratio of two to one and the ratio of thirteen to one. High values are associated with cooler climate systems and in contrast, there are low ratios in grasslands that exist in tropical areas. The total fauna present in the soil is low in the savannas and forests from tropical areas compared to temperate climate ecosystems. The soil fauna is mainly dominated by earthworms but in the tropics, it is dominated by termites which become major components. Interactions of species within the soil appear to be strong as replacements that are competitive occur and effects that are measurable on the dynamics of nutrients but no observations were made on primary productivity (Hiltpold & Rostás, 2017).

Termite and ants can both be species with a strong influence on the dynamics of nutrients and many other effects that are indirect which influence the performance of insects that are above ground. In grasslands that exist in areas with temperate climates, the below- ground standing crops consumed by insects is between two to ten times greater than the mass that is above ground. The effect of insects that are below ground has hardly been observed unless the larvae of scarab beetle destroy grass in areas inhabited by phytophagous nematodes because it heavily reduces in its biomass that isabove the ground. Insects that feed on roots and other species are usually in numbers that are large in the prairies that are mowed and grazed in comparison to those that are not. Herbivores which feed on roots most of the time increase stress to the plant that is hosting thus increasing the number of insects that feed on foliage. Interactions that are complicated take place such as the herbivory suppress root infections by mycorrhozal fungi that can create nematode resistance or insects that feed on roots (Fleisher, 2013).

5.4 GRASSES AND INSECT SUCCESSION

Insects that colonize early successional habitats usually use different strategies. First, like the plant species with small seeds composing the seedsthat create new plant successional species in habitats. Successful insect colonizers include thrips and also aphids that have members which have small bodies such as planktons that live in the air. The colonizers also have bodies that are large with effective fillers with a large range for foraging purposes. Butterflies, ground beetles and wild bees are large species which are commonly known as colonizers.

Frequency, size and distribution of insects within a habitat is newly distributed should be bimodal including the species that is relatively large or small. Grasslands that are young and sown relatively have communities of insects that are less diverse and with more individuals and ratio that is small in relation to predators and preys than grasslands that are old and permanent. Herbivores from the subterranean do not exist in fields that are newly created because they have poor dispersal and abilities for colonizing. With each successional age, there is an increase in the diversity of species (Fleisher, 2013).

In succession that is secondary, annuals dominate in the beginning but are replaced gradually by perennials that are completely superior. The number of insects that bore stems and are specialized, increases with proceeding succession because grasses that are perennial and support many insects go on to replace annuals.

The richness of perennial species increases their attack rates especially by stem borers and it similarly increases with grassland age. The percentage of parasitism can either lead to an increment of attacking stem borers or decrease with each age of succession. Specificity by the hosting plants can vary depending on the insect groups.

Percent macropetry in species that are dimorphic such as plant hoppers that feed on insects is high in habitats that are temporary or successional vegetation that is early, whereas species present in habitats that are persistent are mostly brachypterous. Contrary to the expectations, the fecundity of plant hoppers that are either migratory or diverse is not consistently higher than those of species that are non-migratory because the costs of the ability to migrate affect fertility. Parasitism that is inverse and density -dependent is common in plant hoppers and higher rates of parasitism are most of the times associated with egg densities that are low as found in crops that are agricultural (Fleisher, 2013).

In relation to the conservation of nature, successional fields that have been set aside at the beginning have poor species, a few species of butterflies, bees and beetles in early plant succession of plants they are monophagous. In early successional grasslands, insects also comprise of a few specialists .

5.5 DIRECT AND INDIRECT EFFECTS OF HERBIVORY

Grasses most of the time do not demonstrate a susceptibility to insect attack and they also tolerate grazing pressure this is because the elongation is continuous from the base of the leaf and the bud induction is rapid. Grazing leads to tillering and the conversion of tall canopies into shorter ones and grazed dense lawns. Physiological integration between defoliated and tillers that are less stressed or same gene ramets can be significant in vascular bundle lignification, addition of sclerenchyma in mature leaves and grasses with high silica content can be hard to chew (Hiltpold & Rostás, 2017).

Photosynthesis that is compensatory or stimulated growth by saliva from bovines can play a role in the growth of grass. However, the concept of curves for optimization of herbivores or mutual relationships between grass grazers overestimates the ability for compensation of grasses and is not justified.

For factors that affect grass growth, herbivory is among the stress factors and importance that is relative of every factor is dependent on the context of either the environment or vegetation. Grasses can better tolerate grazing in comparison to many dicots and can, therefore, decrease competition. Grazed areas are inhabited by grasses, other than the fact that most of them are attacked by herbivores than in plants that are well supplied and grow alone. Concepts that emphasize the importance of stress deals with interactions between grass and grazers because release from the competition is usually more important compared to biomass loss caused by herbivores.

Grasses with fungal diseases may decrease the growth of shoots, but if the shoots are infected they are less often attacked through herbivory and grasses may also benefit from such infections. Other examples make an emphasis on the relative stress concept, for example plants that are neighbouring can be competitive and at the same time increase resistance that is associated by acting as a repellant to insects that are herbivorous or also attracting their enemies in a natural manner. When a grass tiller's neighbor is favored less with food for grasshoppers, the tiller experiences an amount of herbivory that can be significantly reduced. Effects that are damaging of the heather leaf beetle lead to movement into grasslands from heathland and poisoning of arthropods changes the ratios of grass and herbs in fields that are successional and this depends on whether foliar or subterranean herbivory is affected. Nutrients that stress plants or water are many at times susceptible to herbivory, but this is not always the case. Large herbivores also produce effects that are direct in the consumption of plants and indirect effects of trampling, burrowing and deposition of urine by insects (Hiltpold & Rostás, 2017).

5.6 FACTORS AFFECTING INSECT POPULATIONS IN GRASSLANDS

The population and density of insect populations is important to the grassland ecosystem and there are mechanisms used to measure this. Climate is one of the factors influencing insect populations. This includes temperature and moisture. It directly affects metabolic processes and the rate of growth but indirectly affects the supply of food and the general conditions of habitats. Weather effects are apparent where climatic conditions are extreme. They have a destabilizing effect on the environment. For many insects, food supply and quality are important factors. The plant varieties that can be exploited by herbivorous plants can be limited by characteristics of plants either physical or chemical. For restricted periods, most insects are able to exploit only a limited range of species of plants and may sometimes may be restricted to certain parts while the hosts may provide food resources for a period of time that is limited.

Interspecific competition among insect invertebrates in grasslands tend to diffuse insect species and therefore appears to have major effects in trends related to population. The intraspecific population ultimately affects the population of insects. Predators, parasites and diseases often cause mortality that are high to which they are responsible for regulating populations of invertebrates in grasslands is questionable. Predators or parasites can exert a degree of control that is significant only within a particular range of populations. In this context, the effects of interaction between the quality of food and enemies found in nature may be of importance.

Human activities will also affect insect population especially in activities such as cutting grazing or burning which are management practices in grasslands. Cutting will affect the population of insect that are dependent on grass species with structures that are complex and reduces the species of grasses that are tall. Species of insects for example beetle are high in grass that is uncut compared to that which is cut. Similarly herbivory can also appear in early succession by insects. Grazing also affects grasslands through feeding on the grass selectively. Heavy stocking of domestic animals will affect insect communities and relaxing of pressure from grazing increases the diversity of species especially those that are phytophagous and parasitoids. Burning is a phenomenon that is less practiced nowadays for the control of pests but sometimes occurs as a natural phenomenon. It leads to the migration of different species of insects and related families in grasslands. Below ground insects are less likely to be affected by a small increase in temperature. Attraction of insects to such areas will depend on the regrowth that occurs after the grasslands burn.

Changes that are modest in relation to the rates of survival or birth or time use in generation induced by differences that are small in nutrition can cause a change that is large in the equilibrium levels of population. According to Stilling,1988, there is no single factor affecting insect population or density that is of overriding importance since different sources of mortality will vary in different ecosystems. The most common key factors were those acting from the next trophic level up such as predators, parasites and viral or fungalinfections. Other factors that were commonly observed to be affecting population density include dispersal, migration and fecundity that is reduced. In half of the studies done, the population of insects can be compared to a building and it can be seen as fluctuating between an upper ceiling and a lower floor rather than the population being maintained around an equilibrium that is fixed. Density dependence was also detected as coming from mortality sources acting from the trophic level below ,for example, from the host plant than from the trophic level that is above. Any given type of mortality also acted to upset population densities away from the mean level than to be dependent on density hence acting to return the populations to a level of equilibrium. The interpretation in relation to the presence of density dependence was disputed by Hassell in the following year as in the presence of density dependence, the numbers of generations available for analysis increased with the same hence a conclusion was made that many of the studies that were done were too short to allow conclusions that were reliable to be made about the occurrence of the dependence on density

A concept that is relevant in the interpretation of population dynamics is the spreading of risks of at least some groups of invertebrates. A study done in the Netherlands on the population of carabid beetles over a long period are subject to local extinctions. The persistence of populations results from dispersal and refounding of local populations. Overall, the survival of the population is dependent on the extent to which there is fluctuation in the local populations out of phase thus leading to the spread of the risk of extinction of a species.

5.7 INSECT CONSERVATION IN GRASSLANDS

Floral diversity can be used to predict the insect diversity in grassland ecosystems. The diversity ofinsect species such as butterflies, wild bees and phytophagous beetles was found to be related positively to the variety in plant species. The age of the habitat and the size of the fragment are known to enhance the number of species in a disproportional manner. The number of parasitoids and predators also increase with the age of the habitat.One of the major threats to biodiversity is the destruction and fragmentation of habitats. Species that are prone to extinction include species of high trophic levels, species that are rear, those with specific habitat requirements, species with populations that fluctuate greatly and species with poor abilities of dispersal. For example, communities of butterflies on grasslands that are calcareous show relationships of species to area that are positive. And species that are endangered benefit from fragments of grasslands that are large. Morphological characteristics in a few species of butterflies for example those associated with flight have shown change with the isolation of habitat fragments with limestone. This shows that habitat fragmentation in human-dominated landscapes that are simple may have consequences to evolution for traits of life history within populations. Changes in community structure and interspecific interactions may lead to disruption in plant and pollinator interactions. Grasses are mostly pollinated by the wind but most herbs play an important role in nutrient poor or grasslands that are managed extensively. Pollinating bee populations need nectar and pollen resources ,as well as, suitable nesting sites. The above may be limited in small grassland fragments and so very small patches receive visits from pollinators. In small calcareous grasslands, the efficiency of pollination, gene flow of pollen dispersal and seed set are reduced.

Fragmentation of habitats also affects populations that are specialized in trophic levels that are higher for example in a plant-herbivore- parasitoid food chain. A community of monophagous butterflies shows an increase that is steeper with the area of grassland than with plant communities. The habitat quality of species- rich grasslands mainly is dependent on the opposing forces of succession and management. The succession speed may be related to the size of the fragment because late-successional trees or shrubs invade from the edge. Grasslands that are abandoned often increase in the species of both plants and insects but decrease in late-successional grasslands. Many butterflies (specialized) occur on regularly mown and grazed calcareous grasslands as they are seen to appear on microclimates that are warm and sparse host plants. Ground nesting species such as solitary bees are similarly abundant in regularly mowed or grazed grass because sparse vegetation and soil that is open provides nesting grounds therefore enhancing their population. The population of above- ground solitary bees is enhanced by dense, high and vegetation that is woody as it offers necessary plant material for nesting purposes. Although a species will benefit from early, mid or late-successional stages thatdepend on their history of life and diversity which is overall high, their conservation should be with a combination of the different stages of succession.

6

Soil Habitats for Insects

CONTENTS

6.1 INTRODUCTION

It is no secret that insects are the most diverse group of organisms on planet Earth and the actual number of species of insects are way more in number than any other group of living fauna. There are currently more than 900,000 (closer to 1 million) different kinds of insects across the planet and there used to be even more but this representation covers approximately 80% of the world's species. Some scientists and experts even speculate that there are even more insects that have not been named by science than they are that have actually been previously named. This figure is estimated to be around 2 million and some extended this up towards of 30 million different insect species that have yet to be described. Within the last decade or even more a lot of scientific attention has been focused towards and entomofauna that inhabit canopies and areas of tropical rainforests across the globe. A good example is the Smithsonian Institute department of entomology that focus on Latin American forest canopies where studies are being carried out which estimate the current number of living species of insects to be about 30 million. In terms of biomass which is essentially descriptive of the total quality or weight of organisms in a given volume of area insects are also probably the largest of all terrestrial animals as it is estimated there are 10 quintillion individual insects are alive today.

In the United States for example there are just over 90000 described species scientifically but however scientifically non described species are estimated to be around 73000 proving just how diverse and widely ranged the insect phylum is. The largest number of described insect orders is Coleoptera that are 23,700 which are the insect order of beetles. There are 19600 Diptera that is the insect order of flies 17500 hymenoptera that is an insect order of ants bee wasps and such and 11500 Lepidoptera which consists of moths and butterflies.

In relation to the number of individual interest in a given area Alfred, there have been interesting studies conducted along those lines where in North Carolina for example soil samples going down at depth of 5 in calculated that there were approximately 124 million animals per square acre with 90 million of them being nights 28 million them being springtails and 4.5 million consisting of all sorts of other insects. In Pennsylvania similar study was conducted that showed 425 million animals per square Acre with 209 million - 119 million springtails and the rest consisting of other different kinds of arthropods. Even specific insect species when studied individually have been found to be quite numerous like wireworms for example which is the lava that click beetles produce being around 3 to 25 million per square acre. The study gets even more interesting when observing insects that live a More Social lifestyle in large nest or swans like termite having nests of more than three million individuals and swarms of locusts holding up to 1 billion individuals.

When specifically studying insects and their habitats it is important to understand first how the Taxonomy of insects work so that we can be able to identify different species and classify them accordingly this of course involves a series of hierarchical categories. Of course insects are animals and because they're segmented with an exoskeleton they fit under the phylum of improper de together with crustaceans. Insects have their own class under insecta and the Order after that is what differentiates them and the five most diverse insect orders have pretty much already been listed as Coleoptera (beetles) hymenoptera (ant wasps and bees) Lepidoptera (butterflies and moths) Diptera (flies) and Hemiptera (true bugs and cicadas).

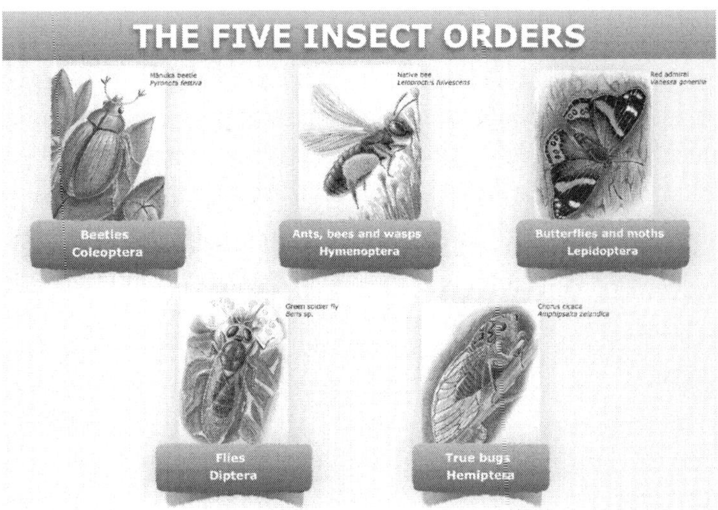

Figure 6.1: Insect orders as stated above.

Source https://www.sciencelearn.org.nz/resources/2840-insect-taxonomy

In essence we can just classify insects as animals that have segmented bodies with six jointed legs and have an exoskeleton. Some of them have wings but not all of them. Some other anthropod such scorpion, spiders and millipedes can be confused and insects however they are not. Insects are one of the most versatile living animals on the planet because they live in almost every environment that exists from the highest Peaks of the world in Nepal to the mounds of the Sahara and all across different tropical rainforests. We will specifically look at what types of insects make their homes and where.

Some insects live in trees for example, the Weaver ant mainly constructs its own homes from natural blue with leaves and other tree material. They work as a team bending leaves with other ants grabbing the original ant and making pocket sheets building entire list also using the Silk that is produced by their young larvae as a type of glue so that they can secure their structures to protect them from predators. There are insects like honey bees build very intricate hives from wax to the produce organizing them into neat hexagons for efficient honey storage. They do this by going out and collecting nectar from flowers which some of they will turn into the Honey that we know but the other is regurgitating and made into a pliable substance that they form these hives out of periods can be built in trees on the branches or inside their Hollows and sometimes on rock crevices. Wasps and hornets also make similar hives but don't actually consume nectar to make honey the same way but have other methods. Some insects are actually aquatic in nature as well and enjoy a life on the water like the caddisfly that is actually a winged insect but lives majority of its life in stagnant or slow-moving freshwater with some even living in salt water as well but this section will focus on insects that enjoy the life on or in the ground and make the soil their habitat.

6.2 THE SOIL AS A HABITAT

Habitat in itself has a various different definition but we can think of it as an area of the environment that provides the necessary resources during the life of the insect in the specific case even though this is a general way of describing it from the perspective of insect needs. Habitation would ordinarily have physical factors like soil moisture different sects of light intensity and a range of temperatures. There are biotic factors that would involve how much food is available in that specific habitat including the variety of predators that share that same environment. Habitat also has geographical types of bipolar temperate subtropical and tropical on the terrestrial ground with different types of vegetation like Forest grasslands semi-arid or deserts and Aquatic habitats including Marshes Rivers lakes and oceans. Geological events like volcanoes tsunamis wildfires and earthquakes make habitats more overtime and they ordinarily have weather patterns that are different as well that also bring different kinds and levels of precipitation and solar radiation (Eisenbeis & Wichard, 2012).

The soil is very special as a habitat and it is a mixture of organic matter minerals gas liquids and different kinds of organisms that all together work to support life. The let's outermost layer that embodies the soil and is subject to soil formation processes is known as the pedosphere which serves four main intrinsic functions;

- It acts as a medium for flora to grow
- It is a container for water storage supply and purification
- It is a modifier of the adjacent atmosphere of the Earth
- Most importantly for this chapter, it serves as a habitat for various kinds of organisms

The soil and its functions serve almost as a symbiotic relationship with the soil modifying the atmosphere for example or the organisms that inhabit it and they in turn modify the soil and its properties. The pedosphere interacts with the lithosphere that is directly beneath it characterized as the outermost shell of terrestrial planets. The song also engages with the hydrosphere that are the combined masses of water on under or above the surface of the planet the atmosphere which is the layer of gases that's around the planet and different biospheres or equals spheres that is total sum of ecosystems around the planet. The soil is studied by scientists as three-state systems of solids that is the minerals and organic matter that make up the soil matrix liquids that are soil solutions such as water inside next to the soil and gases as the soil atmosphere. It can be envisioned as a product of several different factors due to climate elevation orientation and slope of terrain organisms and the soils parent materials are original materials that altogether interact over time. Scientists study soil science are the two main branches that are edaphology which Focuses on the influence of and symbiosis between soil and living things and dollar g that looks at soil is its natural state including its morphology classification and their state of being in the natural environment. When studying soil as a habitat for insects it will fall under the scientific study of edaphology where will look at environmental soil science specifically (Eisenbeis & Wichard, 2012).

Almost a quarter of everything that lives on the planet uses the soil as a habitat living and thriving on or above and below for some parts or their entire lives, depending on it for air water and nutrients. Insects in specific leaves just about everywhere so very hot deserts to super cold snowy -mountain and 97% of them live on land.

Figure 6.2: Soil habitats are everywhere, even in cold dry places like Antarctica. Bacteria and even some insects live in the soil.

Source: https://www.sciencelearn.org.nz/resources/892-soil-is-a-habitat

It can be argued that insects generally are directly or indirectly dependent on the soil as it is the basis of all life. There is a vast number of insects directly associated with the soil and that is why it was important to look at a general taxonomy because the orders of insects from the primitive Thy sanura to the very much more advanced and specialized Hymenoptera indicate widely that all of them are represented in the soil fauna (Collins, 2012).

6.3 THYSANURA AND COLLEMBOLA IN THE SOIL

Thysanura that is the order of insects consisting of silverfish are named as such because of their bristletails and they're in the world over 5 different families 60 general and 320 species. They inhabit the world in a variety of different settings on the bark leaf litter and some species live in caves while others are found in ant and termite nests which are ordinarily in the soil as we will discuss in detail much later on but they have the preference of living in situations where there is high humidity and they feed on starchy materials, for example, wallpaper blue book bindings and even photographs because they are generally vegetarian. They move around quite a bit because they have the ability to jump and they are more ordinarily than not nocturnal. They are most especially abundant in the soil and Collembola more so (Eisenbeis & Wichard, 2012).

Collembola form the largest of the three lineages of modern hexapods which sometimes are debated in the scientific world as no longer being insects and in this case we will discuss them briefly. Springtails or habitually found in the soil with some even more than 6 feet Underground. Springtails normally found in Leaf litter and other decaying material and they are micro B virus in nature and the presence of them in soil proves that the soil is healthy. The fact that they do not bite means that they are not harmful to any of your garden material or plants and the diet consists of decaying matter fungus bacteria and such that provides more nitrogen for the soil.

They are so widely spread that many believe that in terms of sheer numbers they are one of the most abundant of all macroscopic animals on Earth existing everywhere what soil is exposed together with any detailed material. The estimates that say that 100000 individuals of this species exist on every square meter of exposed grounds and other related in habitants like Moss cushions grass tufts and termite nests. In fact, in the insect world only mites are likely to have global populations along with the same magnitude. They are extremely sensitive to desiccation or any form of extreme dryness due to their physiological structure of respiration although some especially anything and permeable cuticles are resistant to drought because they are able to regulate the osmotic pressure of their body fluids (Gill, McSorley, & Branham, 2011).

In terms of the soil itself and how they would distribute themselves horizontally, springtails species are known to have preferences for soil characteristics such as acidity, moisture and light available to the area. There are moisture requirements for some that would explain why they do not live above the ground and retreat well into the soil especially during dry seasons with some found most commonly around ponds lakes and other water sources. It can be found that most soil-dwelling species are not specialized and can be found on most everywhere moving towards more wet areas especially, when laying eggs or when undergoing the process of a crisis which is when the species would shut the exoskeleton or moult.

6.4 ORTHOPTERA AS SOIL INHABITING INSECTS

Orthoptera is an order of the insects that consists of grasshoppers, locusts and crickets and other morphological is similar relatives is divided further into two suborders Caelifera that's made up of grasshoppers and locusts together with their relatives and Ensifera that consist of crickets Katydids and other close relatives. They are an order of insects that undergo what is known as incomplete metamorphosis where the animal has three distinct stages the egg the length and imago that is the adult stage. They are more than 20000 species of Orthoptera that exist worldwide and most of them are able to produce sound known as stridulation by rubbing their wings against each other or their legs which contains some corrugated bumps. They do this because they are social insects and the vibrations help them locate other individuals. Characteristically they are found above the ground and occupy exclusively open habitats and enjoy distributed grassland ecosystems because their diet is vegetarian. They do not mind anthropogenic habitats either and they are important for the ecosystem as a stimulator of plant growth and also participate in the nutrient cycle with major roles being played by them in food chains if the numbers are kept to appropriate levels (Gill, McSorley, & Branham, 2011).

Many of them especially, when their populations grow to catastrophic levels can become devastating for agriculturalists as most commonly witnessed in the East African region during the last several months with the locusta migratoria. They exist abundantlyacross all continents except Antarctica, especially, the Grasshopper species and they are said to cost the world community over half a billion US dollars in damage when they're outbreaks occur affecting ecological processes including the carbon and water cycles. In relation to the soil, they can also affect it negatively as the rapid loss of vegetation subsequently leads tosoil erosion and increases the likelihood of runoff overall affecting biodiversity. In this way we can see that Orthoptera and the soil have a direct relationship

Figure 6.3: Smart drones to be tested in battle against East Africa locust swarms which have been the worst in decades.

Source: https://www.weforum.org/agenda/2020/02/smart-drones-africa-locust-swarms

Acrididae for example which is a family of predominantly 10000 grasshoppers of the 11000 species in the sub order lay there are eggs in the soil usually in a tough outer casing called an Egg pod where a group of eggs are placed within. from here they go through winter and eventually due to the increased temperature and humidity that comes during the spring in terms of rainfall they hatch and the names appear hours vary from lava which are essentially maggot or wireworm shaped with no legs.

When these exit the larva is mostly ectothermic meaning of the relay on external sources of heat to regulate their body temperature where they would, for example, go out in Baskin trucks moving in and out of shade to dilate or constrict their peripheral blood vessels.

Common name	Scientific name	Egg-laying site
Field Grasshopper	*Chorthippus brunneus*	Soil - upper surface
Meadow Grasshopper	*Chorthippus parallelus*	Soil - upper surface
Heath Grasshopper	*Chorthippus vagans*	Soil - upper surface
Rufous Grasshopper	*Gomphocerippus rufus*	Soil - upper surface
Mottled Grasshopper	*Myrmeleotettix maculatus*	Soil - upper surface
Woodland Grasshopper	*Omocestus rufipes*	Soil - upper surface
Stripe-winged Grasshopper	*Stenobothrus lineatus*	Soil - above grass roots
Lesser Marsh Grasshopper	*Chorthippus albomarginatus*	Vegetation - base of grass blades
Large Marsh Grasshopper	*Stethophyma grossum*	Vegetation - base of grass blades
Common Green Grasshopper	*Omocestus viridulus*	Vegetation - base of grass blades

Figure 6.4: Showing the preferred egg-laying sites for UK short-horn grasshoppers (acridids).

Source: https://www.orthoptera.org.uk/about_orthoptera

Phasmatodea are also a suborder of insects that fall under Orthoptera and their relationship with the soil is a habitat also begins from a young age. They are an order of insects whose members consist of stick insect stick bugs walking sticks or bug sticks and they are found all across the world except the Antarctic and Patagonia regions with over 300 species in the island of Borneo thatis the richest place in the world for Phasmatodea. When they lay their eggs ordinarily flicked to the surface of the soil with some of them even burying them in small pits inside the soil. Because many species lay eggs with a fatty cup known as a capitulum they end up attracting ants because they share a resemblance to elaiosome which is a fatty layer that covers many seeds of many plant species. The ant larvae feed on elaiosome and so they would carry the Phasmatodea egg to the nest which is usually underground and remove the capitulum layer that surrounds it to feed their young without actually damaging the phasmid embryo. This beautiful relationship ends with the egg eventually hatching and the young onesactually kind of resemble an ant to leave the nest to climb onto the nearest tree for safety and food in the foliage (a classic example of mimicry in the animal world.) Relatives do grasshoppers and locusts at the gryllidae that conflicts the family of insects related to crickets. They spend a lot of the time on the ground or even underground having a wide almost cosmopolitan distribution in the existence. The greatest diversity of that occurs in tropical and temperate regions like Malaysia where 88 different species can be heard chirping doing nocturnal activities. There are many that inhabit the ground and some even create a subterranean home by excavating shallow or deep burrows. They do this almost as biological defense mechanism because they are relatively defenseless and need to spend the day well-hidden or underground in cracks that are developed in dry weather. Some can even dig the ground folding their antenna to conceal their presence with others having burrows as temporary shelters or even permanent ones where they form residences for mating and laying eggs. Crickets generally are able to dig soil by using their mandibles and carry the loosened soil with their limbs flicking it backwards and pushing forward their head with their Hind legs. The most common of these species that use the soil as their habitat are the ground cricket sscientifically known as the Nemobiinae that are generally when less compared to the other relatives that may have short wings or even two pairs of wings. The diet here consists of seedlings, grass pieces of leaves, and the shoots of young plants however they can be of omnivorous with some that are captive even able to rely on a diet of powdered dry dog food with aphids and Lettuce (Campos-Herrera, El-Borai, & Duncan, 2015).

Figure 6.5: Picture of a mole cricket that lives underground as viewed from a cross section of the soil.

Source: https://www.ardeaprints.com/dec2014/2/mole-cricket-underground-cross-section-earth-10527345.html

6.5 TERMITOIDAE AS INHABITANTS OF THE SOIL

Termites and one of the most successful insects on the planet and they have colonized almost all land except Antarctica generally. They are eusocial insects the colonies having little as a few hundred individuals two enormous amounts of several millions. Termite Queens is extraordinary in the fact that they have the longest life span of any known insect with some Queens living up to even 50 years. They live through an incomplete metamorphosis proceeding through the egg named directly to the adult stage and some species spend their entire lives in the ground. There are maybe 3000 species of termites that are currently classified and over 1000 of them are found in Africa with certain areas having a great abundance of mounds. They have soft cuticles and better control their body temperature they mostly live in tropical regions and dislike cool or cold habitat. Ecologically, there are three groups of termites which are damp wood, dry wood, and subterranean. The damp wood termites inhabit only coniferous forests and the dry wood termites are found in hardwood forests. Subterranean termites live underground in very widely diverse areas (Collins, 2012).

Most of them are very small less than a centimeter in length but the largest being the Queen and measure up to 10 centimeters. The majority of them would be completely blind and don't even have a pair of eyes using other chemoreceptors like the clypeus to make their way around. You will find that males who end up eventually being drones for the Colony form from unfertilized eggs while females that consist of the majority of individuals in the Colony and end up being both workers and the queen form from diploid fertilized eggs.

Young eggs turn into nymphs that closely resemble the adults but are just smaller versions and they go through several molting stages maybe three or four to turn into adults. The nymph will first moult into a worker and from there some will go to further molding stages and become soldiers or atales. This process can take several months depending on the availability of food the temperature in the environment and the overall population of the colony. Nymphs cannot normally feed themselves and so require the workers to feed them but the workers also live a very busy life having other tasks to accomplish such as foraging, building the nest and maintaining it or tending to the Queen. The pheromones that all the animals reduce in the colony will control how the classsystem is organized also preventing all but very few of them from becoming fertile Queens.

The soil mound that builds up the termite nest can be considered as being made up of two parts: the animate and in inanimate parts. The termites would not live in the inanimate side while the animate side is the section that is made up of the living quarters. Termite mounds can be divided into three different kinds;

- Subterranean (mounds that are completely below the ground
- Epigeal (mounds that protrude well above the soil surfaces)
- Arboreal (mounds that are built above the ground but have some shelter tube that connect it directly to the ground)

Most termites actually will have underground colonies that serve as both nests and mounds but other primitive termites will build themselves around wooded structures like logs and stumps. To build these nest termites would primarily use their feces which have proven to be a very desirable construction material because of its properties but other materials include plant digested material that is partly digested, bits of soil and bits of wood are also used. Most nests especially in tropical rainforests are primarily located underground to protect the soft cuticles of termites and help them regulate their temperature. They also serve as a great fortification against predators to protect the termites from light desecration and pathogens (Collins, 2012).

Others that are made out of cotton are usually quite weak so the termites that inhabit them have some counter-attacking strategies that the use against predators that try to invade. Arboreal nests usually are used by mangrove swamp nasutitermes. These nests are normally built from word that has partially degraded and other materials like stems and branches from the adjacent mangrove trees namely the red mangrove the black mangrove and the white mangrove. Some species of termites build and inhabit very complex next known as polycalic nests. These habitats are known as polycalisms. Termites that build polycalic nests form multiple calies that are ordinarily connected by subterranean chambers and they appear to be at least less frequent in mound-building species although polycalic arboreal Nissan also been examined in several species of nasutitermes.

Figure 6.6: A pair of Cathedral mounds in the Northern Territory, Australia.

Source: https://en.wikipedia.org/wiki/Termite#Distribution_and_diversity

Mounds are created by nests that protrude from the earth's surface and they generally provide the same protection as subterranean nests although they are built much stronger. The reason termites build mounds is to avoid the torrential rain that would flood the underground caverns there are more at risk of erosion due to their clay rich construction. There are also built in such a way that they have strong points in case of a breach for example the cubitermes good morning Lee build narrow tunnels that they use our strong points where the diameter of these tunnels just small enough for the soldier termites to block. There is also a very highly protected chamber known as The Queen's sell for the Queen's chamber that house is the King and the queen and is normally the last line of defense (Weissteiner, & Schütz, 2006).

In the insect world macrotermes build arguably the most complex structures in the insect world that consist of gigantic mounds :the largest in the world reach around 8 or 9m above the surface and consist of pinnacles ridges and various chimneys. *Amitermes meridionalis* is just another species of termites that build enormous nests ways ways the largest ever been recorded reaching a height of 12.8 hi and several meters wide and the Democratic Republic of Congo. Marlins are also normally compass oriented North-South so they can avoid the heat of the sun as a form of thermoregulation and some have very elaborate and distinctive forms as shown below.

There are shelter tubes also that termite construct called person tubes or March teams that are found along the walls and other structures. These are normally constructed when the humidity is high and it is a little bit cooler so ordinarily at night. These tubes are used to provide protection from predators especially, ants and they also allow workers to collect food they cannot be accessed in other ways. They are normally constructed from a mixture of soil and feces giving it a brown color and the size of these shelter tubes can be as big as a few meters in length (Weissteiner, & Schütz, 2006).

Figure 6.7: A series of Mounds of "compass" or "magnetic" termites (Amitermes) oriented north-south, thereby avoiding mid-day heat.

Source: https://en.wikipedia.org/wiki/Termite#Distribution_and_diversity

Figure 6.8: Termites in a mound, Analamazoatra Reserve, Madagascar.

Source: https://en.wikipedia.org/wiki/Termite#/media/File:Termites_in_a_mound.jpg

6.6 EPHEMERIDAE AS SOIL INHABITANTS

Ephemeridae are a family of mayflies found almost all across the world except in Oceania and Australia that can have either two or three long tails and can also be quite large up to 35mm with distinctive wings that are elaborately patterned. They like to breed around water and require a layer of silt because their nymphs love the soil and they are built with very strong legs specifically adapted to burrowing and that is why the group is sometimes known as burrowing mayflies. The reason why this is very interesting as well is that they actually have a part to play in the Lakeside of waterside ecosystems changing the water column turbidity its nutrients concentration and even possibly its oxygen balance through their burrowing and ventilation activities. The council burrowing activities of the names also allow soil nutrients to be available for algae to grow that provide asort of food source as well.

Figure 6.9: Female adult (imago) *Hexagenia limbata* (Serville) in Winnipeg, Manitoba, Canada. Photograph by Miles Zhang, University of Florida.

Source: http://entnemdept.ufl.edu/creatures/AQUATIC/mayfly.html

6.7 THYSANOPTERA AS SOIL INHABITANTS

Thysanoptera or Thrips are slender insects that have fringed wings which are generally unsuitable for flight and move quite clumsily in the air. Adults and pupa Thrips will normally burrow into the soil over the winter will come out in spring: for the females to insert their eggs into the tissues of flowers leaves or other stems. Because they live around the plants but they're always consuming it can also be said that they correlate with the soil. Thrips sometimes are considered to be pest especially to commercially important crops and some of them are even vectors to over 20 different viruses that plants develop.

6.8 HEMIPTERA AS INHABITANTS OF THE SOIL

Hemiptera are known as true bugs and include everything from cicadas, water bugs, plant bugs, and bed bugs. They consist of a group of insects of 75000 species worldwide and are characterized by long antenna with more spots that are designed to pierce and suck juices from plants or animals. They exist in almost if not all Forest strata in the soil,leaf litter understory, and over story and play an important role in the deposition of acid in soil Forest ecosystems and are considered to be major soil engineers as well.

They do this by creating abiotic factors that regulate soil microbial activity because the microbes that exist in the soil end up catalyzing biogeochemical reactions leading to the exchange of nutrients between heterotrophic and autotrophic organism. The fact that they acidify the soil is not necessarily a good thing because this can cause toxicity in the trees.

Cicadae for example are adapted insects which the adult versions feed on plants however the name are mostly subterranean inhabitants and feed more exclusively on the xylem sap from the roots of their host plants. Studies have been done where a large number of nymphs of different sizes have been obtained from the soil to show that they are mostly beneficial to the prune mature trees and aerate the soil. If they died before maturing they also serve as an important source of nitrogen for growing trees and some can spend up to 17 years depending on the species in the larval form beneath the soil. In fact, once the emerge from the soil they leave for only approximately 5 to 6 weeks after that meaning that majority if not all of their lives are underground (Eisenbeis & Wichard, 2012).

Aphidiodea or aphids also fall under the insect order of Hemiptera and includes the greenfly and black fly: considered to be one of the most destructive insect pests on the planet especially, in temperate regions sucking the sap from plants and being vectors for plant viruses reproducing rapidly since they are asexual. They seem to share an interesting mutualism with ants as Queens that leave the colonies to start a new one would take an aphid egg to form a new herd of underground aphids. Many even though they live on plants mostly feeding on them would drop the ground for protection and some species use oil to feed on the roots and remain underground their entire lives. There are root -feeding species of Aphidiodea that live a very similar subterranean lifestyle as the Cicadidae discussed above.

Coccidae are scale insects that include mealybugs and consist of more than 7500 species worldwide but were initially called coccids. They are known to be notorious pests that cost 5 billion dollars annually chewing on plant matter from the roots underground to the leaves and flowers above. Nymphs more especially, are found in hyaline spherical cysts in the soil. Although a morphology where is greatly it has been found that many families with some species are produced sources can live underground for many years while some are highly modified to live under the bark of their host plants(Eisenbeis & Wichard, 2012).

Intertidal mudflats which are areas where sediments have been deposited by tides or rivers are also a host to many different kinds of insects including various Galiduidae and Saldidae. Not only are they an important food source for larger animals, but insects breed there in small pools and then larvae are able to feed on various microorganisms such as zooplankton and algae. Cydnidae are a family of Shield bugs that are commonly referred to as burrowing bugs distinguished by ear piercing and sucking mouthparts and their wing composition. In the rainy season, they can be observed in parts of Brazil where flocks are found in the soil although they are generally viewed as pests.

6.9 NEUROPTERA SOIL HABITATS

Neuroptera are the insect order of net winged insects that include lacewings multiplies ant lions and the various other relatives. The most evident and maybe the most striking swell group inhabitants in this order is in the family Myrmeleontidae is colloquially known as ant lions that are a group of insects consisting of around 2000 species known for their fiercely predatory habits especially in the larva stage of life where they dig up pits to trap passerby ants and several other prey. In fact, the adult version of the insects are much less known because they have such a short lifespan in comparison to the lava and they are nocturnal and easily on mistaken in identity to dragonflies or damselflies. And flies are sometimes known as doodlebugs because they leave strange marks on the sand as they set the traps for their prey. They are well -distributed across the globe with a few species that can be found in cold temperatures like the Euroleon Nostras in Europe however, you will find the most commonly occurring in dry Sandy habitats where the larvae can easily dig up their pets where some also known to hide under the bridge toambush the predators around leaf litters (Campos-Herrera, El-Borai, & Duncan, 2015).

As stated above in the larva form, they primarily eat other small anthropods especially, ants and the trap building species dig a pit of about 5 cm deep and 7 to 8 cm wide. There are other insect families under the order Diptera that have been observed to dig up the same sort of pets to feed on ants. Once the individual has marked a sight they build a circular groove around it and they start to crawl backward digging in with their abdomen as a sort of plowor hoe to shovel up the soil. Their front legs have been adapted to this design and they're very strong please sing heaps of loose and particles upon their head and then with a very intelligent technique the jerk and threw a little pile just outside the periphery of operations. After this, continuously it works itself around the circumference towards the center of its trap as it moves round and round concentrically it gets deeper and deeper towards the pit until the slope reaches a critical angle of repose which is the steepest angle that the animalcan maintain where it is just at the verge of collapse with the smallest touch or disturbance. The ant-lion builds the traps and pits with such a technique that at the very edge you will find it is lined with the finest of sand grains. They are also adapted to build it with such efficiency that they minimize the time needed to complete the pit by digging in a spiral manner (Campos-Herrera, El-Borai, & Duncan, 2015).

Figure 6.10: An Ant lion larva extracted from sand pit. Note its very wide build, its strong pincer jaws for subduing prey and its very well modified and adapted forelegs.

Source: https://en.wikipedia.org/wiki/Antlion#/media/File:Antlion_larva.jpg

Once the ant lion is finished constructing its speed it retreats and settles down at the bottom buried well within the soil with only its jaws protruding above the surface wide open in position on either side of the very tip of the visible cone but underground. We steep angle that the strap is built with guides to pray all the way down to the bottom directly into the Lovers mouth while it avoids the craters debris as it avalanches. It is one of the simplest yet most efficient traps in the animal kingdom. Because the ant-lion builds the sloping end of its trap with the finest of grains of sand it ensures that the Avalanche will carry prey that are even as large as possible because it is set up at the angle of repose on the verge of collapse. This provides an insecure foothold for any insect that mistakenly pass over the edge such as, ants. They slide to the bottom and are seized by the lurking ant lion. The trap efficiency is further increased by the ant lion chucking up loose sand from the bottom of the pit causing the prey to collapse further into the trap sometimes actually striking the prey with sand showers.

Ant lion larvae arecapable of subduing even small spiders, a variety of different insects and anthropoids. They have doors that are hollow allowing them to suck fluids out of the victims and after total consumption. The larvae then reset the trap once again by throwing out the collapsed debris from the center of Earth pit steepening the walls again back to the angle of repose. Ant lion larvae require loose soil which is often sent but not necessarily so any large granular material will work for the purposes of Pit constructions and dryer places how much more preferable because they are protected from the rain. When the larva first hatches from its egg, itisa little smaller and so specializes in trapping small insects but as they grow larger they subsequently construct bigger pitsand catch even larger prey sometimes even much bigger than itself. It seems that the animal kingdom has realized the effectiveness of the ant lion and so many other anthropoids actually make use of its ability to trap prey for example, the larva of the Australian horse fly that leaves deep in the pit and feed on the prey that is caught or the female chalcid wasp that that deliberately gets caught in the pittoparasite on the ant lion by ovipositing on the ant lion between its head and its thorax.

Figure 6.11: Sand pit trap of an ant lion.

Source: https://en.wikipedia.org/wiki/Antlion#/media/File:Antlion_trap.jpg

6.10 TRICOPTERA AS SOIL INHABITING INSECTS

Trichoptera also known as the caddisfly, is a group of insects that primarily live as aquatic larva but terrestrial adults. The adult version of these kinds of insects will in some cases use soil for the production of larvae and pupal cases. We cannot discuss the match because even as adults they do not live very long-surviving only maybe a few weeks being very vulnerable to fish as easy pickings and some even do not feed as adults and die soon after they breed but they are important to mention under soil-inhabiting insects because of how they lay their eggs. The larvae of many species when they hatch from the eggs will make protective cases made of silk reinforced with gravel and soil including other debris. They are aquatic and also use the soil and other particles to anchor down to the bottom of the water where they're able to feed an accumulation of silt that is formed when suspended particles are deposited. This behavior is not only studied in science but transfers into the artistic world with the French artist Hubert Duprat who noted their behavior made art by providing them with precious stones and other materials to encapsulate themselves building a unique form of artwork that were food across the world.

Figure 6.12: Hubert Duprat is a French artist known for his unusual work, an artistic intersection between caddisfly larvae and gold, opal, turquoise and other precious stones.

Source: https://en.wikipedia.org/wiki/Hubert_Duprat

6.11 LEPIDOPTERA AS INHABITANTS OF THE SOIL

Lepidoptera is the insect order consisting of Butterflies and Moths with over 180000 species that are currently prescribed in 126 different families and 46 superfamilies. Some of them especially, the females that exist specially in the basil moth's family have true legs that are adapted to boring and living in the sand on every segment of their body in the caterpillar form. These types of insects are holometabolous meaning they go through a complete metamorphosis from the egg to larva to pupa and finally to the imago or adult. The caterpillar form which will be more and discussion especially in relation to insects and soil habitats is the larva stage of this insect and many are built to live in the ground. As the grow the larva changes in appearance going through a series of stages called instars which are general developmental stages between each ecdysis or molting period.

Certain species like the notodontidae and the geomentrina have their pupation periods underground with the crambidae living in well-constructed nests made up of bits of earth. It is commonly known that many larvae once fully matured enter a pupa stage where they spin asilk case or cocoon to pupate but some members go underground to go through this stage. A butterfly pupa is known as a chrysalis that has a very tough hard skin and they don't have a cocoon. Eventually, they emerge from this as a full adult completely metamorphosing and becoming sexually mature.

6.12 DIPTERA AS SOIL INHABITANTS

Diptera are true flies in the insect order and can we argued are ecologically considerably important two human beings and their existence. There exist about 1 million species including horse flies, crane flies, hoover flies but only 125000 can be definitively said to have been classified scientifically. Quite a large number of larvae and pupae of insects ofthe Diptera order are found in the soil usually that which is rich in organic matter. Love is only found in groups that are dependent on food sources that are accessible for only a very short period of time. Some like the tsetse fly will have will have an adenotrophic viviparity characterized by a single fertilized egg being retained in the oviduct and the developing larva feeding on secretions that are released by various glands. Once fully grown, the female finds a soft spot of soil and the larva then works its way out of the oviduct and directly buries itself in the soil to begin its pupation stage.

Although Diptera are known to have wings to fly very close to the ground but there are a few adults that fly to serious heights or even up to 2000 feet over long distances.

6.13 COLEOPTERA AS INHABITANTS OF THE SOIL

Coleoptera is the insect group that consists of all types of beetles. It actually represents the largest orders of insects by far and they are very abundant on and in the ground with roughly 400 thousand species that makes up 40% of all insect species described up to date. They are found in nearly all habitats including freshwater bodies, coastal habitats, and wherever vegetative foliage is found from trees on their bark flowers and leaves to underground near roots. They are seen be as a beneficial resource in relation to human economics by controlling the overall population of pests by feeding on aphids, mealybugs and scale insects. In fact,beetlesgo so far there is no more food sources are scarce they will go out to eat caterpillars and honeydew or young plant bugs. Ground beetles are great in controlling weeds because they eat their seeds while they're still in the soil reducing the need for herbicides to protect crops.

Dung beetles probably one of the most famous,have been very successful in reducing the population of pestilent flies which are a bad pest ofthe cattle in Australia. They do this by collecting the dung which isa breeding ground for the pests and they roll it around and bury it in the soil that has a further added effect of improving soil quality fertility tilt and nutrient cycling. The dung beetle was so successful in Australia that it was actually introduced to South Africa and various other part of Europe to reduce the populations of pestilent flies and following the successful trials the technique was employed even as far as Hawaii. The American Institute of Biological sciences hasshown through a report that they made the dung beetles tosave the United States cattle industry of 318 million dollars on a continual annual basis true bearing livestock feces(Collins, 2012).

Scarab beetles fall under the family of Scarabaeidae and they consist of over 30000 species of beetles worldwide and they spend their egg,larval, pupal and part of the adult stage in the ground. They have bright metallic colors that are attractive with their c-shaped larvae that are called grubs being pale yellow or sometimes even almost white living underground on the debris to avoid exposure to sunlight. They are mostly scavengers that feast on recycled dung, carrion and decaying plant material. *Scarabaeidae elateridae lampyridae* and their close relatives of numerous other species and families are well represented in the soil fauna.

6.14 HYMENOPTERA AS SOIL INHABITANTS

Hymenoptera is one of the largest orders of insects behind Coleoptera consisting of insects : sawflies, wasps, bees ,and ants. There over 150000 known species of Hymenoptera that are being described with 2000 extinct onesand countless others yet to be scientifically listed. Probably the most important soil inhabitants to talk about in this order of insects would be ants that are engineers of entire ecosystems that greatly affect the physical-chemical and biological properties of the soil that their habit. In actual fact, there has been a debate about the existing ants with recent studies showing that the predecessors of early primitive ants rose tens of millions of years earlier than what was previously thought up to maybe 168 million years ago which was just after the rise of diversifying very quickly and assuming ecological dominance maybe a hundred million years after that which is very impressive for any species to have done. They likely went underneath the surface of the soil to escape predators making them the most successful to occupy pretty much all the surfaces on planet Earth. There are however not found in all such as Greenland Iceland and some parts of Polynesia and also the Hawaiian Islands thatall lack native and species. They are eusocial, similar to termitessometimes is called white ants although termites are not and as described above. And different greatly in genetics especially when looking at the reproduction cycle but they have similarities in social structures due to convergent evolution which is the convergence of especially in ancestral groups (Collins, 2012).

Spending the whole lives on or in the ground in a eusocial society means that ants need a communication method which they do have with the use of pheromones sounds and touch. Some groups have pheromones and chemical signals that are more developed than others for example Samantha able to actually perceives smell with a long thin and movable antennas and since most ants live on the ground they're able to leave pheromone trails for other ants to follow for example, when they go foraging in groups. Pheromones are used by ants for more than just making trails they can be sent as alarms to nearby and for attack frenzies that attract.

Ants and various other Hymenoptera membersbuild nest constructions that can be quite elaborate as termites but some species can be nomadic and do not build permanent structures. Nest construction does not differ much from the complex nests that many termite species andcan be found in the ground under stones or logs inside logs only stems or even simply on acorns. The way ants behave is a little different in the sense that they use different materials and may be permanent but, we can say that they are beneficial in the aeration of soils like Weaver ants in Southern China are known to be one of the oldest applications of biological control.

The Formicina are some of the most common ants in places like Australia and can be found in very large numbers foraging on the ground or and vegetation , the soil o, rotten wood and a few are arboreal. Many of them and even other Hymenoptera like the Sphecina and Apina excavate extensive tunnels ,burrowsor cavities where they go in and store their food or lay their eggs and rear their young. Many of the adult species of the Scoliidae can be found in the soil as well burrowing through it and parasitizing on various other insects. Mason bees and wasps this essential material for the construction of their nests well above the ground. A great variety of species in the Hymenoptera order on account of diversity cannot be discussed in detail here and this order makes it one of the most important soil groups (Campos-Herrera, El-Borai, & Duncan, 2015).

6.15 ABUNDANCE OF INSECTS IN THE SOIL

This subject may not be investigated extensively and offers an intrinsically interesting field for research. In 1907 ,McAtee was able to make a consensus on animal life in four square feet of soil all the way down to the death basically that a bird could be able to scratch and feed off of and on this basis he estimated that there are about 1.2 million animals primarily in the in sector order in 1 acre of woodland in about 30 million individual insects in meadowland. On the top of this he was able to find that the word tangible remains of nearly half a million dead insects per square acre of woodland.

In the 1920s, another scientist by the name of Morris in England, Cheshire to be exact, stated that the census of insects found in various samples of soil give an insect population of 3.5 million individuals per acre. The family best represented in this census was the Bibionidae making up around 33% of the individuals. Mycetophilidae made up 17% with Straphylinidae making up 12% of the individuals in the samples studied. When looking at the different number of species, Coleoptera were the most diverse to nobody's surprise with 29 different varieties in that order.

Other studies done within that period include one done by Marlatt in 1907 recorded 84 cicada openings in per square foot of the ground with an estimate of 30,000 to 40,000 in the ground surrounding a fairly decent sized tree. In areas where ants and termite exist, the numbers of individuals present can be very enormous and it might be said that soils that are rich in organic matter ordinary contain thousands of *Thysanura collembola* and various immature Diptera.

6.16 USES OF SOIL BY INSECTS

The uses the Insects make of the soil around them and the benefits that they derive from the soil are almost as numerous and varied as the number of insects that exist in themselves. Of course as like stated above insects are said to utilize soil for shelter ,protection and materials for the various escapades. Some use it as a source of food, moisture, air ,heat, and as an avenue to travel. . Some insectsprefer subterranean territory to avoid excessive climatic conditions or grow through their metamorphic processes. Besides that, however, we have also seen that insects in many instances spend a majority of their entire life is not all of it in the soil like the termites that we have observed excavating large subterranean cavities while some others wood construct just a single cell like bees or wasps do periods these instances would show that some insects build shelters in the soil simply for one or two stages of their life cycle moving on to other territories afterward. The Acrididae for example will deposit the eggs in the soil and all the other stages of their lives after that would occur well above the ground while the Noctuidae can be found in the ground during the larval and pupalstages. In the line of insects using the soil as protection, some insects actually have developed very peculiar markings that confirmed with the general ground color camouflaging them from their enemies. This is very evident in insects in the orders:Acrididae and cicindelidae (Collins, 2012).

Many insects may not live in the soil but use it to construct the different habitats for example Mason bees and wasps can have very elaborate structures built out of soil for laying their eggs and rearing their young. Some Tricoptera may live the water however, you will find that they will construct larvae and pupal cases in the sand. Some carbide for example will lay eggs and place them in cells made of soil and mud before attaching them to plants.

The soil has served insects as a vast storehouse for food but various challenges can occur when it comes to studying the feeding habits of subterranean insects. You will find that many of them are free- living: moving from place to place when looking for food while others are sedentary and some leave the soil environment entirely returning when they are searching for food material period and for example will store seeds grow entire fungus bushes and take care of aphids. Members of Carabidae and Asilidae will rummage through the soil searching for prey likewe've already discussed. The ant lion constructs conical pit to trap their prey. Solitary wasps will stuff their excavations with insects and spiders while solitary bees will fill theirs with pollen. *Canthon laevis* will construct balls of manual sometimes bigger than themselves which they carried to a suitable place and then bury it while *Corpis carolina* packs its Burrow using it (Collins, 2012).

6.17 MODIFICATION AND ADAPTATIONS FOR LIVING IN THE SOIL

Insects are of terrestrial origin and have developed in such a way to fit into the terrestrial habitat and the fact is emphasized by primitive insects more closely related to that same environment. Practically all orders and nearly all the families represented in soil fauna have simpler forms of metamorphosis especially with majorly subterranean groups. Some interesting modifications may be important to mention like some insects that have shown great and striking development of their interior legs mainly for borrowing and digging up the soil to transport it. The mole cricket for example, has well- modified front legs that resemble a mole that is very well adapted for digging up the dirt and shoving it aside and also for cutting off the roots of plants. The anterior legs of the cicada that live an incredible 17 years display an interesting enlargement of the femur and tibia that somehow resemble the cutting mandibles of biting insects. Their anterior tarsus goes through metamorphosis during the adolescent life where they modify according to the activities that the insect carries out (Campos-Herrera, El-Borai, & Duncan, 2015).

In conclusion, you will find that insects utilize the soil in numerous ways and in many cases they are dependent upon it for their very existence and in some way there exists a sort of symbiotic and reciprocal relationship since insects are an important factor in soil formation, renovation and ,maintenance. This fact is majorly overlooked by very many entomologists but is well- observed and recognized by soil students and scientists. The importance of insects in the soil becomes significant in exception to protozoa and nematodes, insects constitute the largest group of soil fauna.

Many insects like we have studied bury plant and animal matter in the soil, feeding it and making it rich in nutrients and minerals. Some would even go as far as ingesting the soil whereby they are acted upon by the various insect secretions causing beneficial chemical changes. This insect-soil association is fundamental to the development of life as is on Earth

7

Forests Habitat for Insects

CONTENTS

7.1 INTRODUCTION

An area where theclimatic conditions support the growth of trees is known as a forest. Forests cover small to large areas. For instance, tropical and deciduous forests cover few square kilometers with a wide variety of tree species. Other forests cover large areas inhabited by fairly few species. Forests comprise of abiotic components such as landforms, soil, water and air as well as biotic components like trees, animals, plants. According to the Food Agriculture organization-FAO (2015), globally forests cover over 4.06 billion hectares which are about 31 per cent of land and has approximately three trillion trees.

Roughly, half the forest land is virgin (an area that shows no or little human activity). Worldwide, forests are unequally distributed. A big percentage (66%) of forests grows in ten countries. When we talk about forests, we think about the following five countries where about half of the forests are found.

Forests contribute a lot to plant and animal survival without which there would be no life. Majority of the world's terrestrial diversity (80%) lives in forests. People depend on forests for fuel, water, building posts and fruits. Communities that live in or near forest areas depend on trees for their livelihood security. Forests provide fuel wood, food, and fodder for domesticated animals. Forest dwellers harvest poles to build houses, medicinal herbs to treat diseases. According to the FAO, 2018, approximately 880 million people spend a big portion of their time producing charcoal and collecting fuelwood. Ninety per cent of the people, who live in dire poverty, depend on the forest for livelihood security.

Trees play an important part in the formation of rain. They release water vapour into the atmosphere through transpiration which rises and changes into water droplets that form clouds. Once the clouds are saturated they fall as rain. Trees protect soils from the harsh impact of raindrops by sieving them before they hit the ground. This reduces soil runoff and promotes soil percolation. Water infiltration into the soil helps the absorption by plant roots hence keep the plants well-watered. Trees trap poisonous gases such as carbon dioxide (Act as carbon sinks) and produce large quantities of oxygen that is needed by all living organisms. They act as windbreakers that isthey reduce the speed of the wind and protect the crops on the farms. The leaves that the trees shed provide humus and organic matter that is rich in soil nutrients crucial for plant growth. Numerous living organisms find primary forests ideal since they are serene places where they go about their business undisturbed. They compete for available resources and finally strike a code on how to live with one another. Ecology has shown that animals that live in the forest have developed food chains, for instance, a zebra feeds on the grass while lions prey zebras. Just like humans, birds build their nests on the branches of trees, some animals and birds live in the hollows, insects and animals in the forests live in communities and homes. As one walks across the forest, they cannot but help admire communities of monkeys swinging from one tree to another. Grass- eating animals such as zebras, antelopes and buffaloes live together. Foxes dig holes and live there while most birds are always busy building nests. They add to the aesthetic value of forests and that is the reason why tourists find their way there. Entomologists have witnessed that forests provide good breeding grounds different classes of insects.

7.2 FOREST LAYERS

According to World Atlas, forests naturally arrange themselves vertically in different layers which include emergent, canopy, understory and undergrowth. Trees form strata based on their heights. Different forest types have different strata. Some temperate countries in temperate region such as Germany and Poland use a different classification of tree, shrub, herb and moss layers.

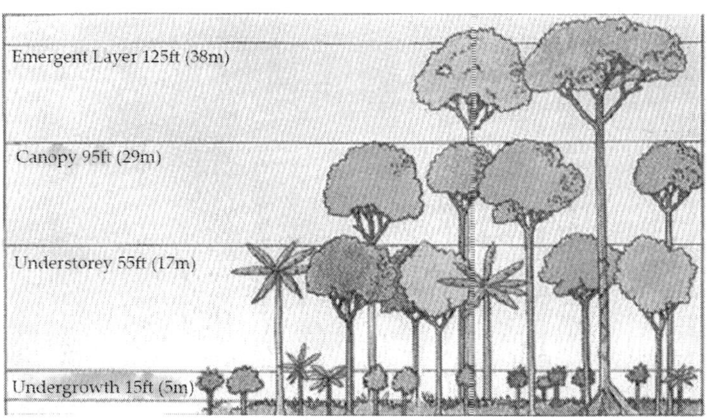

Figure 7.1: A diagram showing forest layers; Source ZME Science

The emergent is the topmost layer comprises of trees, woody climbers, and epiphytes. In this stratum, trees often tower to between seventy to eighty meters high. This layer features in tropical forests only.

The canopy layer comprises of inter-twined branches, twigs forming a shady, protective "roof" over the rest of the forest.Trees in both temperate and tropical forests form the canopy. The majority of the trees that form the canopy grow up to sixty meters while the average is around twenty meters. In the temperate forests, trees in this layer grow up to ninety meters. The canopy of the tropical forests intercepts about ninety-five per cent of sunlight enabling the trees to perform photosynthesis rapidly supporting production in these forests. Flowers and fruits develop, mature and ripen fast. The canopy additionally, reduces the force of rainfall. Raindrops hit the leaves of the trees reducing their impact in the ground hence reducing soil erosion. The canopy shields forest from strong winds and storms leading to reduced soil erosion. The common animals that live in this layer include monkeys, birds and leaf-eating insects (Einspruch, 2013).

The understory layer comprises of trees that require little sunlight and wind. It is vertically located below the canopy. Trees in this layer are scattered as opposed to those that form the canopy. The trees stand at ten meters in height. The forest canopy prevents the sunlight from penetrating the trees below since the trees that grow. Trees in the shrub layer need even lesser sunlight than those of the understory layer. Shrub comprises of young trees and bushes which is further subdivided into upper and lower bushes. Generally, it is hard for the shrubs to tower above five meters. Shrubs are common in deciduous temperate forests. They are rare in primary tropical forests but can be found in areas where original tropical forests have been cleared. Since the bushes are thick, some bird species reside in this layer. A few examples of shrubs include elder, hazel, hawthorn, raspberry, blackberry and clematis. Shrubs protect the soil from losing moisture.

The forest floor or sometimes referred to as a non-woody herbaceous ground cover. Compared to the upper strata, this layer has very few species. In the original tropical forests, this layer has more species in temperate forests and covers larger areas as opposed to original tropical forests which cover less than ten per cent of the forest area.

Tree leaves and animal droppings fall and decompose on the shallow soil layer where moss, cryptogam grow. The decomposing organic matter and cool temperatures are favourable for cryptogams the species that propagate through spores such as ferns, mosses, lichens, fungi and algae. Mosses grow to about 0.15 meters. Decaying leaves and logs provide a good environment for the growth and development of insects and other small living organisms. These organisms breakdown the matter into the soil and provide nutrients to plants (Einspruch, 2013).

7.3 FOREST TYPES

Forests thrive in different climatic regions. The main forest regions are classified according to their proximity to the Equator. The tropical forests are found within the tropics (near the Equator), temperate forests are common in the temperate region farther away from the Equator than the tropics and boreal forests (taiga) are found in the polar region. There are other sub-categories within the major types. Forests are also said to be either evergreen or deciduous. They either have broad or thin leaves (Coniferous) or mixed.

The forests that grow around the Equator are called tropical rain forests. These forests are common in Africa, South America and Southeast Asia. They have a wide variety of diverse species per area. They cover a small area yet half of all species-about 1.4 trillion. Tropical forests receive rainfall year-round (at least 200cm per year). The temperatures in the forests are around 27 degrees Centigrade. Aside from the rainy seasons' tropical forests have dry seasons. The Equatorial climate supports the growth of a variety of plant species. According to ZME Science, a quarter kilometre of a tropical forest house about 100 tree species. Trees grow densely and the leaves and branches block sunlight penetration into the understory. Some of the species include broad leaf trees, mosses, ferns, palms, and orchids. Some of the animal species that are found in the tropics include frogs, lizards, monkeys, and snakes (Einspruch, 2013).

Figure 7.2: Tropical rain forest: Source ZME Science

The soils in the tropical forests are thin due to percolation of nutrients. The topsoil which contains decaying matter supports plant growth in the forest. The sub-types of tropical rain forests include, an evergreen which experiences no dry season, and rain all the year-round, seasonalones have a short dry season and evergreen vegetation, long dry season characterize the dry category.

Characteristic of the montane forest: Montane forests are mainly fed by mist and fog. They mainly confer. The tropical and subtropical coniferous type receives a dry and warm climate with conifers adapted to variable weather. Sub-Tropical forests are adapted to counter summer drought and are found in north and south of tropical rain forests proper.

Temperate forests are found in Northeastern Asia, North America and Europe. They encompass four seasons including winter. Temperate forests receive 75-150cm of precipitation throughout the year. Temperatures range from negative thirty degrees Centigrade to positive thirty degrees Centigrade. The main trees species include deciduous trees (that sheds leaves) and coniferous trees as well as firs and pines. The average tree species is three to four trees per square kilometers. Some tree species include beech, oak and maple. The following animal species thrive in these forests rabbits, squirrels, birds, deer, bears, wolves, and foxes (Einspruch, 2013).

Temperate evergreen coniferous forests are common in, New Zealand, and Northwestern Europe, north-western America and South Japan. Trees in the temperate zone is roughly 0.6 trillion. They bear the name temperate from the fact that they receive high rainfall of between one hundred and thirty to five hundred centimeters per year. The forests are dominated by confers. The tree species that are commonly found in this region are pine, spruce, cedar, cypress, fir and redwood. Deciduous trees such as maples and many mosses and ferns are also found here. Animals that live in these forests include bears, owls, deer, elk, and marmots.

Figure 7.3: Coastal temperate rainforest. Image credits: Sam Beebe

The sub-types in the temperate rain forests include moist conifer and evergreen broad-leaved, dry conifer, Mediterranean, temperate broad-leaved rainforest. The tree species common in this region are evergreen broad-leave moist conifers. It experiences mild wet winters and dry summers. Dry conifers are common in higher altitudes which receives little rainfall. Almost all the Mediterranean trees are evergreen and grow along the coast, south of the temperate regions. Temperate broad-leaved rainforests receive heavy rainfall throughout the year as well as mild, frost-free winters.

Boreal forests also referred to as "Taiga" are found in the Sub-Arctic zone between fifty and sixty degrees latitude. Boreal forests cover Siberia, Scandinavian, Alaska and Canada countries.

This zone has a long cold dry winter and a short, moist, mildly-warm summer seasons. Temperatures range from as low as negative forty to twenty degrees Centigrade. This region receives between forty and a hundred centimeters precipitation per year. These areas have a thin layer of acidic soils which are low in nutrients. Not much light penetrates through the canopy hence little growth in the understory layer. Needle-like evergreen trees that tolerate cold temperatures grow in this region, for instance, fir, pine and spruce are the main trees species that grow here. Likewise, the animals that thrive in this region cope with long periods of cold temperatures. Some of the animals that survive the cold weather include bears, moose, lynx, wolf, deer, wolverines, caribou, bats, small mammals, and birds.

Figure 7.4: A boreal forest. Image credits: Cephas

Other types of forests include plantations and natural forests. Plantations or artificial forests are planted and usually of the same species such as cypress, pine and eucalyptus while natural ones are indigenous and have a wide variety species. Natural forests are also referred to as virgin forest that a human has had limited or no influence.

7.4 FOREST GROWTH

Forests appear when temperatures rise to ten degrees centigrade and the annual precipitation is over 200 millimeters. The type of soils, plants, and animal life vary according to environmental conditions. As seen above, hardy conifers such as pines, spruces, and larches grow in high-latitude subpolar regions.Conifers and broad-leaved deciduous trees dominate the temperate high-latitude climates. Deciduous broad-leaved forests grow in middle-latitude climates which receive an average temperature of above ten degrees centigrade for at least six months every year. The annual precipitation is above 400 mm. Within a hundred to two hundred daysdeciduous forests oaks, elms, birches, maples, beeches, and aspens grow (Einspruch, 2013).

The Equatorial climate supports the growth of tropical forests.Heavy rainfall allowsthe growth of evergreen broad-leaved vegetation. The temperate deciduous forest reappears in the lower latitudes of the Southern Hemisphere.

Forest types are identified by species composition, the density of tree cover, type of soils found there, and the geologic history of the forest region.Soil depth determines the extent to which roots can penetrate the earth and, the amount of water and nutrients available to the trees.

For instance, areas where taiga forests grow, the soilis sandy thatdrains quickly. The soils that support the growth of deciduous forests are brown, richer in nutrients, and less porous than sandy soils. A soil layer rich in iron or aluminum giving soils either a reddish or yellowish allow rainforests and savanna woodlands to grow.

The amount of rainfall received in a region determines thewater moisture in the soil. The water available in the soilsdictates the type of trees that grow. Water from the ground through evaporation and leaves by transpiration forms water vapour. Temperatures in the forests are determined by this evaporation and transpiration. In the cold season, the temperatures in the area around the forest are warmer. The areas surrounding the forests are cooler during the warm season (Einspruch, 2013).

The amount of both sunlight and rainfall that reach every forest layer is determined by the density of tree cover. Between sixty and ninety per cent of available light is absorbed by the leaves for the photosynthesis in a full-canopied forest. The leaf cover influences the movement of rainfall into the forest. They reduce the impact of raindrops which penetrates down to the ground level by running down tree trunks. The water that is not absorbed by the tree roots for nutrition runs over which under normal circumstances in forests is minimal.

7.5 WHAT IS A FOREST HABITAT?

The place where plants and other living organisms live and get their basic needs such as water, food, shelter, space for survival, protection and mates for reproduction is commonly referred to as a habitat. Habitats are either biotic or abiotic. Biotic consists of living things such as the presence of predators and availability of food while abiotic refers to non-living things like soil, moisture, temperature and light intensity. Just like man living organisms have homes (New, 2011).

Every organism has conditions under which they thrive though some live in a variety of environments. Termites live in holes, birds live in nests. Habitats vary from one organism to the other and the need for survival, for instance, the pitcher plant is found in boggy areas that have plenty of moisture and receive ample sunlight. They require soils low in pH level. Bearing in mind these specifications, there are very few areas where the pitcher pant can thrive and are endangered by habitat destruction.

The interior of a stem, a rotten log, a rock or a clump of moss can be a habitat forinsect larvae and adults. Parasites live in the hosts' body. There is different type of habitats. Geographic habitats are distinguished by different zones including, polar, temperate, subtropical and tropical. As the name denotes, terrestrial vegetation habitats are different types of vegetation which include, forests, steppes, grasslands, semi-arid or deserts. Fresh-water habitats include marshes, streams, rivers, lakes, and ponds. Salt marshes, the coast, the intertidal zone, estuaries, reefs, bays, the open sea, the sea bed, deep water and submarine vents make up marine habitats.

As time elapses, habitats morph. Changes are caused by natural phenomena such as the eruption of a volcano, an earthquake, a tsunami, a wildfire or a change in oceanic currents. Climatic variation over time may cause these changes slowly. For instance, the recession of glaciers on Mount Kenya as a result of global warming turned into deserts. Damming increases rain in surrounding areas. Others result from human activities, such as diverting and damming rivers, deforestation, the ploughing of ancient grasslands, the draining of marshland and the dredging of the seabed. The introduction of foreign species causes an imbalance in the ecosystem. Species compete for resources and some species become extinct while others overpopulate to dominate the area. Scientists use the introduction of alien species to eradicate harmful species from farms. For instance, they introduced the lady bird to get rid of aphid pests (Leather, 2008).

Another classification of habitats includes polar, temperate, tropical and subtropical habitats. Polar habitats are located on the northern polar region (Arctic) and the southern polar region (Antarctica). The polar region is windy and cold. Tundra is an example of a polar habitat in the arctic region. The ground is almost always frozen. Some plants such as grasses and mosses grow in summer Polar bears, Arctic foxes and wolves are among the animals that thrive in this region.

Temperate habitats situated between the subarctic and subtropical zones. Temperate habitats are located in the temperate regions. The seasons in this region includesummer, autumn, winter, and spring. Temperate forests are found in Eastern North America, Northeastern Asia, Western Europe, and Eastern Europe. Some of the animals that live in this zone include the eastern chipmunks, white-tailed deer, and European robins.

Subtropical habitats located between the tropical and the temperate zones in the subtropics. Palms are common in this region. Insects are found in all kinds of environments. Some live in the desert, temperate, or the tropics (Leather, 2008).

7.6 INSECTS IN FORESTS

Insects and trees have co-evolved over time and eradicate each other. Individual insects being fragile and sensitive to changes in the environment, they reproduce highly to compensate for the high death rates. When trees change their habitats due to wildfires, hurricanes, landslides, a shift in climatic conditions, insects follow them.

The role of insects in land and freshwater ecosystems cannot be underestimated. They assist with the following processes, pollination, recycling nutrients, sustaining the food chain and keeping forests in check.

In forests, insects pollinate eighty per cent of all the trees. Flies, beetles,bees, wasps, and butterflies suck nectar, eat pollen from flowers and deposit it in other flowers, pollinating plants. They gather food from different sources and in the process, some fall by the wayside and germinate (Szujecki, 2012).

Most of the Hymenoptera members have relatively unspecialized mandibulate mouthparts. An exception is found in the bees (superfamily Apidoidae) where the maxillae and labium are modified into a proboscis that works like a tongue to collect nectar from flowers. In these insects, the mandibles are used to gather or manipulate pollen and wax (Szujecki, 2012).

Ants contribute a lot to the distribution of seeds. Ants stock food for a rainy day.

Termites and wood roaches occupy similar habitats, share the same type of food and provide care for their offspring. Termites are the only hemimetabolous insects that exhibit true social behaviour.

Insects also feed on leaves and needles and in so doing, regulate nutrients and energy sources in the soil. They breakdown the organic matter found on the forest floor to contribute to soil formation. It is easier for plants to absorb nutrients once broken down.

Lepidoptera (moths and butterflies): This is the second largest order in the class Insecta. Almost all Lepidopteran larvae are called caterpillars. The majority are herbivores; some species eat foliage, others burrow into stems or roots, and the rest are leaf-miners.

Adults have large wings compared to body size. Their bodies are covered with minute overlapping scales. Lepidopteran wing scales often produce distinctive color patterns that play an important role in courtship and intraspecific recognition.

Brightly coloured, butterflies have hooks at the tip of their antennae. They are diurnal-meaning that they work during the day. When they rest, they hold their wings vertically together. Most moths work at night, therefore, they are nocturnal. They have comb-like antennae. Moths are typically drab in appearance. They hold their wings horizontally when the rest.

Orthoptera (one of the largest plant-feeding insects) are terrestrial, herbivores with modified hind legs that are adapted for jumping. They make and detect sounds.

Aphids have piercing and sucking mouthparts and feed by withdrawing sap from vascular plants. Nymphs live underground and feed on the roots of trees and shrubs.

Earwigs are mostly scavengers or herbivores which are active during the night and hide in dark recesses during the day. Some of them are predatory. They feed on a wide variety of plant or animal matter. After laying eggs in the ground, females have the behaviour of guiding them till they hatch. Females lay their eggs in the soil and may guard them until they hatch. The cerci at the end of the abdomen are enlarged and thickened to form pincers (forceps) which are used in defence, courtship, grooming, and even to help fold the hind wings.

Insects also facilitate the process of decomposing trees:

When a tree falls, insects drill holes into the tree trunk and allows fungi, microorganisms and other insects to find their way into the fallen tree or log and break it down further. Without the insects the microorganisms would find it hard to work, it would take them twice as long to decompose trees (Szujecki, 2012).

Once insects break into the wood, "pioneer insects" such as bark beetles, longhorn beetles, jewel beetles, timber worm beetles and wood wasps assemble around it to feed on the bark and sapwood (the outermost layer of the trunk's wood). The decomposition process begins. Then the twigs fall off and another set of insects, bacteria and fungi work on them. These reduce the rotting wood into thick, nutrient-rich material called humus. Ants, fly maggots, beetles, mites and springtails work on the humus. Worms and snails come to move the decay along to create even more surface area for the microbes to work. This process produces soil humus eventually breaks down, leaving us with soil.

Proturans (Coneheads) are common in moist habitats in temperate deciduous forests.

Diplurans live in moist soil, leaf litter, or humus. They are predators since they feed on other organisms that live in the soil.

The springtails not only feed on decaying vegetation but also soil fungi.

Silverfish feeds on starchy vegetable matter, algae, and lichens. They live under stones.

Flea larvae have a sparse covering of bristles and are worm-like (vermiform) in shape. They scavenge on organic debris (including adult faeces) and rarely attack a host. Compared to fly larvae, flea larvae can survive more arid conditions. Fleas pupate within a thin silken cocoon after molting twice. When conditions allow, adult fleas can live for up to one month.

Bark beetles so to say monitor forest health since they feed on the sick and weak trees eliminating them. They improve the forest resistance and the overall wellbeing. They take part in the decomposition of the deadwood returning its nutrients to the soil. However, an increased bark battle population has a devastating effect on the forests contributing to bad health. This is situation is made worse by external factors such as climate change and wildfire outbreaks.

Insects sustain the food chain by feeding on leaves, wood and bark. In the process, bugs are eaten. Other forest creatures such as woodpeckers, warblers and sparrows, as well as mice, shrews, bats, salamanders, frogs, toads and lizards, depend on bugs as a food source. Some insects eat other insects.

Mantids have long front legs with spines for catching and holding prey, a head that can turn from side to side, and cryptic colouration for hiding in foliage or flowers. They are most abundant and widespread in the tropics.

Coleoptera (beetles and weevils) is the largest order in the class Insecta. They inhabit all terrestrial and fresh-water environments, have a wide variety of diets, and exhibit different lifestyles. Most species are herbivores that are adapted to feeding on the roots, stems, leaves or reproductive structures of their host plants. Some species feed on fungi, others burrow into plant tissues, others excavate tunnels in wood or under the bark. Many beetles are predators. They live in the soil or on vegetation and attack a wide variety of invertebrate hosts. Some beetles are scavengers, feeding primarily on carrion, faecal material, decaying wood, or other dead organic matter. There are even a few parasitic beetles, some are internal parasites of other insects, some invade the nests of ants or termites, and some are external parasites of mammals

Predatory or parasitic insects eat overpopulated or invasive plants or other insects, thus keeping pests or weed populations under control and benefit their ecosystems. Insects also act as a source of food for other organisms especially birds, frogs, and lizards.

Adult fleas are parasites and survive on sucking blood. The majority live on mammals and around 10% depend on birds. Adult fleas cling to the host's body and suck its blood. Most fleas spend a considerable amount of time away from their host, unlike lice. The average life span of the adults is one year or even longer. The blood that a flea sucks can sustain it for weeks or months.

All living beings however small need each other. The part each being plays is vital to other organisms': plants or animals. For instance, forests are the source of water that is a vital commodity for all living beings. They provide clean air by absorbing poisonous gasses emitting into the atmosphere. Insects as we have seen above play a vital role in releasing nutrients into the soils.

Cockroaches also referred to "water bugs", are scavengers or omnivores are common in the tropics. They are also found in the temperate and boreal regions. Cockroaches have an oval, somewhat flattened body that is well-adapted for running and passing through narrow openings. Much of the head and thorax is covered and protected dorsally by a large plate of the exoskeleton.

The Heteroptera (true bugs) include a variety of insects adapted to a broad range of habitats- terrestrial, aquatic and semi-aquatic. Terrestrial species live in plants. They feed on nutrients stored within seeds. The scavengers live in the soil or underground in caves or ant nests. The predators survive on a variety of small arthropods. Others in the Cimicidae family such as bed bugs live exclusively as ectoparasites on birds and mammals (including humans). Aquatic Heteroptera members live on the surface of both fresh and saltwater, near shorelines, or beneath the water surface in nearly all freshwater habitats. With only a few exceptions, these insects are predators of other aquatic organisms.

Herbivory common in the gall wasps, and ants and bees. They are either predatory or parasitic. The greatest diversity, though, is found among the many families of parasitoid wasps whose larvae feed internally on the living tissues of other arthropods (or their eggs). These insects complete their own larval development and eventually kill their host. They have a small size and characteristically narrow host range, these wasps are highly abundant and exert a tremendous impact on the population dynamics of many other insect species.

All Dipteran insects are true flies. The hind wings are reduced to small, club-shaped structures called halteres (only the membranous front wings serve as aerodynamic surfaces). The halteres' vibration helps the insect maintain balance as they fly just like a gyroscope.

The larvae of Dipteran insects have no legs. They reside in moist environments in aquatic (freshwater), semi-aquatic. They are commonly found in the soil, in plant or animal tissues, and in carrion or dung- almost always where there is little danger of desiccation. Some species are herbivores, but most feed on dead organic matter or parasitize other animals, especially vertebrates, molluscs, and other arthropods. Adult flies live in a wide range of habitats and display enormous variation in appearance and lifestyle.

However, when insects attack and destroy healthy trees then they act as pests. Pests are the single most cause of forest destruction each year. Many studies have been conducted on the harmful insects. Humans apply pest control measures thus destroying myriads of insects in the process. Good forest management eradicates such an insect invasion. The likelihood of insects attacking healthy trees is nil. Trees follow a line of weakness. It could be a sickness or even a crack. If the forest invasion is detected in good time, then forest destruction can be avoided.

7.6.1 Defoliators

As the word denotes, defoliators are insects that strip trees of their leaves. Insect larvae in the form of caterpillars feed on tree leaves. They get rid of chlorophyll, the substance that a plant needs to photosynthesize crippling the ability of a plant to produce food and energy. Persistent defoliation can lead to the extinction of tree species. Overpopulation of defoliators can cause havoc. Examples of defoliators include grasshoppers, pine webworm, pine sawflies, gypsy moth, forest tent caterpillar, eastern tent caterpillar, bagworms, cankerworms, beetle larvae and leaf-cutting bees. When defoliation consistently takes place over a long time, then trees die. Defoliation harms more young trees as opposed to mature ones. Coniferous trees are more susceptible to defoliation while deciduous hardwood trees are more tolerant. Defoliation can slow down plant growth. Inserts exert more stress on trees if they attack during dry spells. While some insects conceal themselves in clipped foliage, silken tents, web-woven shelters, folded foliage and leaf tissue others feed in groups covered with toxins. Spruce budworms (a moth) lay their eggs on Douglas fir and true fir and the caterpillars that hatch feed on the needles of these trees hence, weakening them. The caterpillars attach themselves to trees with silk which they spin and if they fall off the trees, ants which live on the forest floor feed on them. Several defoliators are seasonal, for instance, are during winter, others have references (Szujecki, 2012).

All the defoliating caterpillars that considered as pests of forestry trees in South Africa are native. They are also strongly polyphagous (many host plants), where one moth species often feeds on some native and introduced hosts. Infestation by caterpillars can be very sporadic, where heavy infestations in a particular area may only occur every number of years. However, some areas experience high infestations of a particular caterpillar every year. It is only the caterpillar stage that feeds on the leaves (Einspruch, 2013).

Emperor moth larvae are a common defoliating pest on forestry trees. Some of the most common moths include the wattle emperor moth (*Gynanisa maia*; infests Acacia), the pine emperor moth (*Imbrasia cytherea*; infests Pinus and Acacia) and the poplar emperor moth (*Pseudobunea irius*; infests Eucalyptus, Acacia and Pinus). The full-grown caterpillars are very striking and very large with spines and tubercles on their bodies. Pupation occurs under the needle mat in the soil. Colored eye-spots on their hind wings are characteristic of moths.

Euproctis terminalis a pine browntail moth is a major defoliating pest of Pinus. The caterpillars feed on the basal part of the needle letting the remaining part of the needle to fall to the ground. The presence of this insect can be sensed by an abundance of green needles on the forest floor. The caterpillars have poisonous hairs that cause severe irritation on human skin.

An outbreak of these caterpillars can also pose a danger to forestry workers. The pupae are found in the needle mat and are covered in the toxic hairs. The adult moths have a distinct dark brown tuft of hair at the end of the abdomen (Tooke 1938).

Wattle bagworm (*Chaliopsis junodi*) is one of the most serious defoliators of *Acacia mearnsii*. The caterpillars of this moth live inside a bag made from silk and material from its plant host. Pupation takes place inside the bag. The adult female remains grub-like in the bag while male adult moths emerge from the bag (Einspruch, 2013).

Other defoliating caterpillars include the brown lappet moth (*Pachypasa capensis*; infests Acacia, Eucalyptus, Pinus), the pine looper (*Cleora herbuloti*; infests Eucalyptus, Pinus), and the wattle semi-looper (*Achaea lienardi*; infests Acacia) (Web 1974).

7.6.2 Bark Borers

Bark borers insects bore into the bark of trees to feed or reproduce. They destroy the phloem, a thin layer of cells found beneath the surface of tree bark that transports carbohydrates and nutrients throughout the tree. When bark borers destroy the phloem around the stem then they end up killing trees. The most damaging pests in the forests are bark borers. Once bark borers dig into the bark of a tree, they lay eggs that hatch into beetle larvae. The larvae feed on the phloem tissue that is rich in nutrients and grows into an adult. Adult borers fed on tree barks. Strong health can prevent the bark beetles from entering into the bark but can become vulnerable in the event of an outbreak. Some of the bark borers include southern pine beetle, Ips beetles, black turpentine beetle.

Bark beetles reproduce in the living and dead phloem as well as cambium tissues that are found in the inner bark of a tree. Some species, such as *Dendroctonus ponderosae* known as mountain pine beetle can attack and kill live trees. Majority of them live in weakened or dead trees.In original forests, bark beetles act as catalysts here they hasten the recycling and decomposition of dead and dying wood and renewing the forest. The active bark beetles reproduce rapidly, invade and kill healthy trees and are known as pests (Paine, 2007).

Scolytinae bark beetles feed and breed between the bark and the wood of various tree species, including spruces. Over twenty species feed on weakened, dying, or dead spruce, fir, and hemlock. Majority of the bark beetles limit their breeding area to a specific plant part such as the branches, twig, stems and roots. Some breed in a specific tree species as opposed to those which breed in many tree species.

Bark beetles target trees that are weakened by disease, drought, smog, overcrowding, conspecific beetles, or physical damage. As a form of defence, healthy trees produce sap, resin or latex that contains insecticidal and fungicidal compounds that can kill or injure attacking insects, or simply immobilize and suffocate them with the sticky fluid (Paine, 2007).

According to the United States of America Department of Agriculture, in 2005, an outbreak of Mountain Pine beetles killed millions of acres of forest from New Mexico to British Columbia.

Native and introduced beetles defoliate forest trees in South Africa. The native beetles, similar to the defoliating caterpillars, often feed on many native and introduced host plants, and infestations can be very sporadic. The introduced beetles will often have a more limited host range, particularly one the genus of the host plant, and infestations by these beetles are more regular. The latter is partly due to the absence of natural enemies to control populations of the pest.

7.6.3 Wood Borers

These are wood-boring insects that bore into the dead, dying trees or green logs to lay eggs. The larvae that hatch from the egg feed on the inner bark and boreholes deep into the wood. While wood borers cause degradation of lumber but seldom cause tree death. Boring exposes the trees to fungi attack causing discoloration. The larvae can live in the tree bark for a long time only to be discovered if the tree bark is peeled. Examples of wood borers include sawyer beetles, ambrosia beetles and sirex woodwasp.

7.6.4 Seedling, Twig and Bud Pests

These are insects that infest seedlings or the young succulent tissues of small twigs, shoots and buds. They cause death and disfigure large trees. Seedlings can be killed, and larger trees can be disfigured. The activities of these pests can be confused with those of other pests. Some of these pests include tip moths, twig girdlers and white pine weevil.

7.6.5 Piercing/Sucking Insects

Various insects harm trees by attacking the surface of soft plant tissues and feeding on nutrient-rich sap. When many insects invade a particular plant they are likely to deplete the carbohydrates produced from photosynthesis. If infestation continues for a long time, tree deaths are imminent. Some of these insects carry pathogens that can cause a decrease of a tree species or death. The majority of the piercing/sucking insects are among the most destructive forest pests. A few of these insects cause little or no harm at all.

Some of these insects are too minute to be seen with naked eyes and cause minimal damage to trees at a short period. However, they can be a menace over a long time. They can be observed with a magnifying lens. Without this a lens, one can only observe their impact on the plants mainly thin foliage and discoloration of the plants. Examples of these insects include scales, aphids, gall makers, hemlock woolly adelgid and balsam woolly adelgid.

Eucalyptus weevil (*Gonipterus scutellatus*) is a terrible defoliating pest of Eucalyptus. It was first detected in South Africa in 1916 and introduced from Australia. The larvae and adult feed on the leaves. Larvae feed on the epidermis of the leaves whiles the adult eats the whole leaf and the shoot. An outbreak of the Eucalyptus weevils leads to stunted growth of the tree. It lays eggs on the leaves in rows covering them with excreta that form a hard capsule. Previously, this pest was active in high altitude areas, but it has increased in number and spread to lower altitudes.

Another beetle is the Eucalyptus tortoise beetle (*Trachymela tincticollis*) introduced from Australia and first detected in South Africa. Once this pest was detected, a biological control agent was introduced which suppressed it. Though rare, high infestation of this beetle can cause an alarming damage on the Eucalyptus tree.

Green bronze beetles (*Colasposoma* sp.) are brightly coloured beetles that infest Eucalyptus and Pinus. The damage is observed on young trees and transplants. This pest can cause death when it feeds on all the leaves and ring-barks of the plant.

Elliminestes laesicollis a grey weevil attack transplants and young trees and outbreak may lead to stunted growth. It is smaller in size than the Eucalyptus weevil and appears during the replanting of sugarcane.

Additionally, wattle chafer (*Hypopholis sommerii*; infests Acacia, Pinus) and the lesser wattle chafer (*Monochelus calcaratus*; infests Acacia) are considered as minor pests. However, high infestations have been recorded. For both of these insects, the larvae are white grubs. The Catamonus weevil (*Catamonus* sp.) attacks *Acacia mearnsii.*

Exotic insects are one of the worst threats to native forest ecosystems since the tree species and the insects did not co-evolve. The trees lack a natural way of dealing with such attacks.

7.7 DRIVING OUTBREAKS

Climate is a major driver of the outbreaks, according to Hicke. Warm winters help the mountain pine beetles and other bark beetles to survive while cold winters kill them. Warm temperatures favour the growth and development of beetles that sometimes causes a population explosion and thus increasing the destruction of trees.

Trees are more susceptible to beetle attacks during drought and high temperatures. For instance, pinyon pines normally withstand engraver beetles (*Ips confusus*) but they are stressed and susceptible during the dry season. Insects migrate from the very cold to warm environments. Warm climates expose trees to beetle attacks.

7.8 INSECTS IN THE TROPICAL RAIN FORESTS

Insects in the tropics account for a quarter of all the animal species in the world.

Surprisingly, they consume far more herbage than the vertebrate herbivores. Some insects are minute and their presence is noticed in leaves with holes.

Another way that plants interact with insects in the tropical rain forests is as pollinators.In the tropics, insects are the most important pollinators of tropical forest plants. The most important pollinators are moths, butterflies, beetles, flies, and bees. Some insects visit specific flowers while others pollinate a variety of flowers. Through time, complex and fascinating relationships have developed between many tropical forest flowers and their pollinators. For instance, the relationship between the male cuglossine beesand orchids. During pollination, orchids produce fragrant chemicals that prolong the life of the pollinating bees. Pollination is as important as life in plants (Einspruch, 2013).

Some insects in the tropical rain forests are predators, for instance, parasitic wasps. They lay their eggs in or next to a host insect and when the eggs hatch, the larvae depend on the host for survival. The parasite can end up killing the host. The "fairy flies" or Myrmaridae, the smallest insects in the world (under a quarter of a millimeter long) are parasitic on the eggs of other insects. They are very small and hence ignored or unnoticed hence under-recorded. They may even outnumber the beetles. True predators are found in many insect classes, for instance, the lacewings, mantids tiger beetles, robber flies, and assassin bugs. Predators and parasitic wasps regulate the populations of herbivorous insects. Insects may seem to have a negligible contribution to the conservation of ecosystems compared to the big five lions yet they are capable of causing a lot of damage to the tree (Einspruch, 2013).

Insects in the tropical rain forest act as decomposers.Vertebrate dung in the tropical grasslands forms a part of the nutrient recycling process. Appropriate insects help in building up the dung as required. Vertebrate dung is of only minor importance in the tropical forests but is of enormous ecological interest. The adults of the Panamanian sloth spend most of their lives in the canopy and have the peculiar habit of climbing down the tree and hanging from a vine as they defecate. As they do this, they lay their eggs in the fresh dung. In the tropical rain forests, most of the decomposing material comes from plants. These include dead trees, small twigs, leaves and fruits. Termites play an important role in decomposing wood in the rain forests and their impact varies from one forest to another. Other scavengers include beetles, crickets and cockroaches.

Ants are involved in the decomposing cycle. However, the leaf-cutting ants cut and chew up leaves that is used as a medium for the growth of fungi, which is then eaten. Almost every group of terrestrial organisms has developed a form of relationship with ants. For instance, some plants release nutrient-rich secretions to attract ants. The plant then benefits since ants remove the harmful herbivorous insects. The "ant plants" of the tropics which are epiphytes with chambered provide free lodging to ants. As a sign of appreciation, the ants keep off the herbivorous insects and place debris in the plant chambers that become humus for plant growth. In the tropics, many insects have developed complex ways of benefiting from the ant's labour. It is no wonder that one can find beetles, scale insects, lacewings, mites, parasitic ants, flies, wasps, and woodlice in ant nests. They come there to hide or eat the food that the ants have gathered (Einspruch, 2013).

Ants are scavengers and also voracious predators of invertebrates, most notably the infamous army ants of the neotropics. Very few ants are herbivores, the exception being the driver ants of Africa. However, the leaf-cutting ants can be major defoliators of plants. They cut and chew up leaves which are then used as a medium for the growth of fungi, which is then eaten. Tropical forest ants are of enormous interest to all ecologists studying adaptation and evolution. It seems that almost every group of terrestrial organisms has evolved some sort of relationship with ants. Some plants bribe ants onto their leaves with nutrient-rich secretions and subsequently benefit as the ants forage on their leaves for insects and thus remove injurious herbivorous insects. This type of mutualism has reached the extreme with the "ant plants" of the tropics. These plants are epiphytes with chambered tubers that provide the ants with free lodging. In return the ants keep off the herbivorous insects and place debris in the plant chambers. The debris acts as the humus for plant growth. A far different type of symbiotic relationship can be seen with the leaf-cutting ants that have evolved a complex interdependence with the fungus that they culture in their nests. The animal kingdom holds numerous examples of mutualistic and parasitic relationships with ants. Many insects have evolved sophisticated ways of reaping the benefits of the ants' labor. Ant nests in the tropics contain not only ants but also beetles, scale insects, lacewings, mites, parasitic ants, flies, wasps, and woodlice which eat the ants' food (Einspruch, 2013).

In the tropical rain forests, insects play an important part in the food web. They are the main pollinators. Insect predators and parasitoids may regulate the populations thousands of species of potential pests.

7.9 HUMAN INFLUENCE ON FORESTS

Human activities are contributing to deforestation and forest degradation. As the population grows, more and more people clear forests for agriculture and shelter. Humans have generally decreased the amount of forest worldwide. Other human factors that have led to reduced forest cover include logging, urban sprawl, human-induced forest fires, acid rain, invasive species, and the slash and burn practices of swidden agriculture or shifting cultivation. As trees are felled, the insects that live there go down with them. The International Union for the Conservation of Nature and Natural Resources (IUCN) estimates that 420 million hectares of forest have been lost since 1990 through conversion to other land uses. Globally, the area of original forests has decreased by 80 million hectares since 1990. The rate of deforestation between 2015 and 2020was estimated at 10 million hectares per year. A total of 20334 tree species were included in the IUCN Red List of Threatened Species (IUCN, 2019a) as of 2019. Over 1,400 tree species are assessed as critically endangered and in urgent need of conservation (Paine, 2007).

Habitat destruction is a situation in which the natural habitat is unable to support the species rendered functionally unable to support the species present. As more humans move into these areas and invasive plant species are introduced, the habitats of pitcher plants decrease greatly.

Generally, the healthier the forest, the more resistant it is to widespread pest attack. Over mature, weak, wind-thrown, and lightning- or fire-killed trees have little or no defense against infestation and are a factor in the buildup of pest populations.

Selective cutting of susceptible trees, thinning reduce the host material and breeding grounds of pests that could spread to healthy trees. This and other forest management practices stimulate vigorous tree growth and are good methods for indirect control of insects. These practices in regions with a high incidence of a known pest, foresters attempt to avoid serious trouble by planting only trees known to resist existing pests in the regions where the trees are grown. Many forest genetic programs have as a major goal the selection and breeding of trees with insect and disease resistance (Paine, 2007).

Occasionally the natural conditions that suppress the population of pests change, and outbreaks in forests may reach epidemic proportions. Even-aged stands and plantations with trees of the same species and of uniform size and age often create perfect conditions for the rapid spread of insects and diseases. Even the more complex uneven-aged forests with their inherent check-and-balance systems can develop devastating populations of pests. At this point the forester must consider direct control measures.

When the potential mortality is extreme, a direct control of insects of standing timber though expensive is generally employed. Routine monitoring of insects and diseases allows foresters to schedule timely harvests of infested trees and to limit the spread of the problem to uninfected trees or areas. These sanitation and salvage harvests, coupled with piling and burning the limbs and branches left after logging, reduce the material and conditions that allow pest populations to develop. Traps baited with sex-attractant chemicals, or pheromones, are a promising method to reduce the breeding populations of certain insects. To check a sudden plagues of insects, insecticidal sprays from the ground or from low-flying aircraft offer a short-term.This method has been applied to curb defoliating caterpillars, such as the gypsy moth in the United States, the nun moth in central Europe, and the pine looper moth in England.

Pesticides are distributed when caterpillars are most vulnerable. These blanket treatments have a disadvantage since it does not eliminate parasitic and predatory insects that serve as natural controls on the pest's numbers. The chemicals sometimes affect birdlife and contaminate waterways affecting aquatic invertebrates. Large-scale chemical treatments of forests are uncommon and restricted to a small proportion of the areas at risk. Natural control through predatory organisms, which also cycle opportunistically in a slightly delayed sequence with the pest populations, combined with physical factors like cold winters, provides adequate checks. In some situations, biological control involving the release of predators or diseases of pests is promising. Logging transfers beetles from one area another introducing them in areas where they never existed before (Paine, 2007).

In conclusion, forests are home to a wide variety of insects. Forests are suitable environments for them since most of them have little or no disturbances at all compared to agricultural zones. As observed above forests and insect have co-evolved over time so they have developed ways of coping. The trees provide food for the insects while the insects help with the process of returning nutrients to the soils through breaking down dead wood and litter. The insect though minute plays a vital role in the forest ecosystem.

8

Urban Insects

CONTENTS

8.1 INTRODUCTION

The urban condition could be a complex of locales made by individuals from common areas or nations. Houses, towns, urban zones, structures, and various highlights that depict the urban condition that reliably also desolately relocated the trademark as well as nation zones. As a portion of this change, two or three living spaces and their related plant and creature systems were shed, however others were widened plus unused ones were made. Different of new characteristic environmental factors were deliberateness parks, conductors, road trees, turfgrass, food stores yet a couple were indispensable standing water in roadside channel, refuse and landfill objectives close private neighborhoods, the underground sewer and whirlwind direct sort out in urban and rural zones. They all gave regular environmental factors to a select gather of frightful little animals and various arthropods, some of which accomplished difficulty status. Neighborhood conditions, air, and accessible assets choose the spread of two or three arthropods in an urban condition, and for two or three creature classes their riches is restricted. Various species are widely changed as per the advantages and harborages in and around structures, and these are catholic in a difficult situation status.

Reliable locales with assets and conditions legitimate for long stretch continuance bolster store masses of bug species, and from these circumstances people or social affairs move or are moved to set up intrusions in temperamental or brief spaces. Peridomestic districts are found outside as well as the surrounding edges of structures. They join the exterior of structures, the lavish trees, shrubs, and turfgrass which depict the urban and country scene. Nuclear family living spaces are inside, and join the materials that are based on plants and animals in this controlled, anthropogenic condition.

8.2 PERIDOMESTIC

Harborage substrates, edible assets, as well as customary conditions of Urbanized areas all over the globe everything looked at help as a wide number of specific species, in case not singular species in immense numbers. Earth's possessing also settling arthropods in this condition meld ants which scavenge inside and termites which hurt principal wood, wasps, bumble bees, and irregular bugs, for instance, clover bugs, millipedes, centipedes, and springtails. Plant-dealing with terrifying little creatures utilize the made urban plus nation trees as well as bushes, and many are stylish upsets. Blood dealing with vermin, mosquitoes, ticks and gnawing flies are related with private as well as non restrained vertebrates.

Species use surfaces and edges of substrates to join the umbrella wasps, yellow jackets, hornets, bugs, also scorpions. Underground sewer and whirlwind spillage channels give two or three cockroach and rodent species get to urbanized plus natural neighborhoods. The junk move make out of grouping, orchestrating, and landfill give harborage and food to cockroaches, flies, rodents, and bug flying animals. Storehouse people groups for different of the bug species set up in per domestic living spaces are in neighboring typical or undisturbed zones. Forest tracts and green space patches can bolster people groups of hornets wasps and gnawing flies, ticks, and dreadful little creatures. Here are the majority which give people or packages which improve and recharge invasions to little predictable living spaces, and reestablish masses lost to living space demolition (Klotz, 2010).

8.3 DOMESTIC

Biological environments inside stand as commonly steady, also the harborages plus edible assets are to some degree constrained. There may be hardly any species, anyway those changed as per explicit resources consistently happen in expansive numbers. Put aside food, counting packaged whole food and vegetables, ordinary surfaces, and various materials are the main fundamental harborages and food assets in the family unit condition. Direct or by recommendation related to these are dermestid frightening little animals, flour bugs as well as flies, moths, plus cockroaches. A general development in private things and similar storage environments across cultures has contributed to the cosmopolitan bug status of different of these alarming little animals, in both private and business regions. Skin and blood taking care species which breed inside are confined but however, lice, bugs, kissing bugs, are restoratively basic bugs for more than one money related degree of society.

Horrendous little creatures plus various arthropods in living spaces exacerbate upsets when they are very few and their embodiment brief, anyway are not endured when they represent prosperity risk plus hold tight by their far reaching populations. Trademark areas plus people groups for several private creature gatherings, especially those tormenting put aside edible substances, have been lost. Only peoples inside the urban condition address different of these species, or they make due allegorically talking through their interface with people. Some indoor bugs have store masses in peridomestic as well as essential zones. Different fundamental species come up inside homes that feathered creature plus rodents which and from that point approach indoor areas.

The individual fulfillment for the vast majority inside the future is determined by the idea of urban territories. As of 1950, about 60% of the world's kin inhabited towns plus little towns in farmland. Constantly by 2030,it is projected that 60% of the same will live in metropolitan arrives. Those urban regions will be more prominent than at some other time and rule the scene: most by far of these urban zones will be in making nations. Hazardous progression in urban people groups and the determined stream of improvement in individuals from the field put incomprehensible strains on city associations and the individual fulfillment. Lodging, success care, sanitation as well as water systems have to be in touch with unforeseen development, also disease threats. In spite of the conditions, developments to urban communities continue which is incredible news. Urban people group give headway and progression, and generally an improved life than in ordinary compasses. Extended length of many made nations is related with their urban zones (Klotz, 2010).

Urban improvement is sure; the test is the best approach to address the outcomes and push ahead the individual fulfillment from downtown zone to the unconstrained housing at the edge of the city. Frightening little animals and various arthropods that pass on and transmit malady living things demonstrate a peril to the urban zones and thickly populated urban zones of the world. In these territories, amassed everyday conditions and done for sanitation bolster vector people groups, and the centralization of human has can keep up essential infections and quickly spread unused ones. Inconvenience association and control systems will be established on inconvenience perceiving confirmation and life-cycle data, an understanding of deftly conditions, and sensible substance and nonchemical control materials (Klotz, 2010).

8.4 AN URBAN SITUATION

Huge circumstances can be completely appointed trademark, natural, and urban. Common circumstances are rough objectives where the connection plant plus creature systems are still to be changed by human movement. There are relatively few, if any, of these on the planet nowadays, and an undeniably down to earth importance of normal organic frameworks can be undisturbed spaces that have had a limited human impact, and hold a bundle of their striking verdure. A crucial element of these living spaces is the majority of close by plants as well as creatures. These are considered to be the graceful masses of different species that consequently adjusted to cultivating and urban conditions. Agrarian plus urban circumstances are described by their utilization also how much their biotic plus abiotic highlights changed by the human turn of events (Robinson, 2005).

The term urban is a great part of the time utilized reciprocally with city, anyway when utilized inside the setting of the urban condition it loosens up to plant and creature arranges in urban networks and including provincial zones. There's a continuum of included objectives and human activities to metropolitan spot of business from the unrefined farmhouse, and the divide between urban, nation and natural is severely portrayed. The urban condition has levels of adjustment and changes inside the physical scene and biotic systems like those discovered inside the agrarian condition. In urban natural frameworks, we have a wide range of collaborations and blends of the urbanization and specific world and the already existing or introduced provincial or ordinary features or grounds. Coming up next are a few models:

8.4.1 Influence of Agribusiness

The urban interface with building up an extraordinary piece of the time happens when country spread, passing on with it private and business land use, is made near animal farms. Dairy cows, animals plus poultry practices are routinely infringed upon as the characteristic ring of urban systems' spreads. The flies commonly related with the misfortune at these endeavors can scatter a few Km and make an unsettling influence entire months in a year. Dairy steers social occasions could have 50–200 dairy creatures; in mellow territories they are housed in horse refuges for some time in a year; in warm areas they are outside when in doubt. Poultry egg age is regularly talking in 100,000 winged animal structures in social events of in any occasion 10, and they work all year. Excrement passed on at these activities can back gigantic masses of house fly and dependable fly also in a few areas. A traditional laying hen weighing 1.8-kg passes on for all intents and purposes 113g of soggy compost a little bit at a time; normally 11,300kg for reliably or 4,139metric tons for consistently for every 100,000 flying creature poultry building. Feedlots may have 1,000–3,500 dairy creatures in a consistent movement, and every feedlot cow like makes around 23,000g of immersed fecal issue bit by bit. Stable fly and house fly hatchlings require around 2g of manure to mean headway, from this time forward the potential for creatures and poultry endeavors to pass on flies and issues is fundamental. Other feedlot works out, for example, those for turkey and chicken creation, have game plans of feces. Going with the storing of animal and poultry excrement could be a concentrated fertilizer smell, and as a rule too a trouble in the midst of a large portion of the year. Grown-up flies usually cover 20 km and over from recreating zones and at times be given by winning breezes to nonfarm destinations and be an irritation and a portion of the time a flourishing risk (Robinson, 2005).

Urban society in urban scopes could have assortments of animal manure and related flies supporting on this advantage. Zoos, pet inns, corrals, plus massive interruption subject parks have titanic animals, and manure ejection at these territories can be perilous since it may not be reasonably spread on outlining farmland. Fly masses at these objectives might be run of the mill in smooth regions, at any rate little measures of grown-ups will appear in the midst of winter months. Other disturbing little creatures are related to manure and flies, such as yellow coats, carabids, also compost scarabs (Klotz, 2010).

8.4.2 Influence of Trademark Zone

The urban interface with undisturbed or trademark locale happens when nation spread brings private lodging movements near or at the edge of show up put in a sheltered spot or made sure about as a conventional zone. Wild or generally undisturbed zones may give deftly people groups to private and peri local trouble species, checking yellowjackets and skilled worker bumble bees, carpenter ants, underground termites, and several kinds of ticks also vermin. Huge as well as little creature masses would add to capacity of arthropod-borne ailments, for instance, Lyme disease, Cruel Mountain spotted fever, West Nile contamination, also torment. Numerous rugged or wild regions utilized for entertainment in the western US of America contain huge masses of plague-positive rodents. Countless torment cases inside the US of America have been contracted amidst recreational interests, or in provincial districts flanking to wild appear. Expanded urban headway has realized extensive amounts of individuals living in or close domains with rat masses that harbor torment. The peri residential characteristic environmental factors made in private neighborhoods give harborage and food for versatile rat species, for instance, squirrels plus shake squirrels, chipmunks as well as meadow dogs. The amount of confirmed occurrences of torment inside the United States of America plainly transmitted by family cats is developing and is when in doubt related with private spans. These participation's offer course to a huge amount of frightening little animals to take the urban locales that present a positive natural framework for them to call home and breed in. In this way, the coming up of urban normal environmental factors make and are unending until control measures are taken.

The fundamental unpredictability of urban networks combines includes that give harborage plus edible substances for arthropods as well as various animals. Parks, preoccupation zones, and other greenspace have trademark living spaces for vertebrates as well as yellow creatures; arrangement of whirlwind water plus sewer channels gives made conditions for various creatures (Dhang, 2014).

Squander collection focuses and landfills are dependable features of urban conditions far and wide, and these goals give conditions for arthropods, rodents, and bug winged animals.

Trained creatures' farming inside the condition of poultry egg and meat age, feedlots for pig, and meat dairy cows sometimes interface with private and business appear.

Some urban situations are:

Parks, greenspace, and nurseries: Various urban territories have been outlined to meld space for huge and little stops, forested zones along little streams and streams or fringe green belts. These zones break the gruffness of private and business structures, influence neighborhood temperature and dampness, and give neighborhoods with an open redirection territory. In front of timetable inside the improvement of urban territories inside the US of America and Europe sweeping tracts of show up were put something aside for parks. At the point when set up and orchestrates into the scene and ordinary activities, they ended up an essential piece of the urban condition. Urban people group can have two classes of open degrees or greenspace: those that have been intentioned set up as parks or entertainment plots, and the spur of the moment objectives of void parts and roadways. Previously, the shifting attributes of plants and creatures may be kept, and these regions are fairly affected by use instances of individuals and nuclear family pets. Void bundles, porches, roadway center strips, and the benefits of-technique for railroads and various lanes may have a wonderful game plan of plants and creatures. Present day roadway and freeway structures that enter or circle urban districts reliably have wide medians and shoulders, and these may be planted with fences, trees and turfgrass.. These dainty bits of appear a great part of the time have an enormous and unmistakable invertebrate fauna.

Going with the later marvel of ceaseless the suburbs and broadening nation regions has been the expansion in family bloom gardens. Regardless of the conditions of urban raised structures and a strong and dull top substrate, urban nurseries are thriving in different areas.

An urban or nation scene of trees, greenery, or blossoms consolidates money related impetus to property: in two or three cases an expansion of 12–30% can be developed. In any case, the extensive all-inclusiveness of family and open nurseries can too be joined by some prosperity risks being wellbeing dangers. Be it neighborhood or enchanting plant species are used, urban nurseries may give food, condition, or harborage for invertebrate ailment vectors and their vertebrate has. Urban untamed life, for instance, rabbits, deer, chipmunks, mice, plus voles, eating a mix of create plants or seeds, and people groups much of the time become extensive and badly arranged to control or in all actuality direct. Their irritation status relies upon harm to plant plants, settling inclinations, and filling in as has for ticks and other parasitic dreadful little animal vectors. Growthes in Lyme malady plus Horrendous Mountain seen fever in the eastern parts of USA could be credited to deer and rabbit populaces.

Storm sewers and clean sewers: A fundamental urban structure is the sort out of underground channels that reasonable waste water from storm water flood, kitchens and toilets. Different of the storm channels and urban sewers worked inside the 1700s and 1800s are as yet being utilized, and in two or three urban networks they have been expanded or connected with beginning late made systems. This clarify spillage structure is hidden from see, and maybe from the affirmation that it sometimes gives harborage and food for mosquitoes, cockroaches, rodents, and other vertebrate and invertebrate disturbs. The openness of focal points and uniform trademark conditions consistently works out as expected in all year inconvenience people groups in these underground channels. Urban zones may have contrasting frameworks for overseeing family squander water and for exhausting surface or whirlwind water. A joined structure joins family waste and surface overflow water into one organize of channels and gives the solidified release to a concentrated sewage treatment office. A few urban networks have a structure which has family waste and whirlwind water to detached channels. Those channels passing on simply surface water release at different fixations into normal courses, and the waste water is encouraged to a sewage-treatment office. The detached structure includes the greater bit of surface water to storm sewers, yet a few it may be gotten together with sewage and regarded some time starting late being discharged.

While the two techniques can give harborage and various assets for bugs, the joined waste water structure is bound to fortify trouble peoples, since of substances contained inside toilets also kitchen decrease. Whirlwind water spillage structure of channels redirects a ton of water that will something other than what's expected assemble on roads and streets taking after over the top storm or day away from work. Water from streets and walkways streams into the underground mastermind of pipes through sounds and catch bowls orchestrated along the check and city convergences. Channels are ensured about by a pound and identified with a catch bowl some time starting late making a beeline for a drainpipe. A catch bowl is commonly a rectangular cutoff box found underneath the road. It is needed to trap road garbage some time starting late it enters and obstructs the surge of water into drainpipes. It is not up to the point in which water goes to a particular height inside the catch bowl does it stream into the critical storm exhaust. Since of their new development and underground region, get bowls reliably hold water for noteworthy stretches. The mix of normal issue and standing water in a reduce and made sure about zone gives a raising zone for two or three kinds of mosquitoes. These objectives furthermore give a wellspring of food for cockroaches and rodents. Comparative conditions are appear in a segment of the underground mass-travel frameworks and shopping locales in critical urban networks globally (Dhang, 2014).

Strong squander trade and landfills: Grouping and move of strong waste are fundamental to human prosperity and the bit by bit action of a city. Squander passed on by nuclear families and business sources is assembled and exchanged to a landfill, a region gave and particularly administered for squander trade. It may be near the city or passed on to a far off area. Basic strong waste beginnings from reliably improvement in families, lodgings, focuses and thriving consideration workplaces, and burger joints, and it contains 10–half damp and putrescible regular surface. The tall customary substance may be a potential food resource and harborage for startling little animals, bug fowls, and rodents. The utility this surface should these bugs is impacted by the strategies utilized for arrangement, and the brief-and long stretch evacuation. Open deny objectives may be the key method for social event the decline from little systems or neighborhoods in specific bits of the world. These objectives are when in doubt uncovered, three walled canisters, gigantic metal holders, or fundamentally an empty plot of land. Dependent upon the check of the regions served, there may be in any event one of them in a region.

Ignoring the way that this method leaves customary deny defenseless against inconvenience invasion, storing up family deny in apportioned areas empowers successful departure and is unparalleled than uncollected decline inside the street. Dependent upon air and typical temperatures, repeat of combination, and the normal substance, open deny districts support broad invasions of flies and rodents, and reliably pull in winged animals, pooches, cats, goats, and various creatures (Dhang, 2014).

Rodents and flies may set up long stretch people groups at these areas and move starting there to scrounge in or swarm including structures. Fly parasites inside the reject at the hour of arrangement may be expelled from the populace, full-developed hatchlings leave their breaker to pupate and keep up a key decent route from variety and stay to rein fest. Hot and dry environment can lessen the intrigue of deny burdens to flies, and hot and damp climate may heighten it. Energized plastic or steel holders with covers are usually used to hold family dry and putrescible surface. Ideally, the deny is secure until washed down into an arrangement vehicle, anyway best on waste holders may not altogether dodge area of wasps, flows, rodents, plus shocking little animals.

In several urban networks, 60% of the waste holders could be amassed by fly hatchlings. Rodents chomp little openings inside the foot and sides of plastic compartments and take off them available to help snare. The most noteworthy purposes of waste holders are reliably not utilized, and waste is uncovered. Reliably or consistently squander collection is reasonably work of air and the nearby professionals. Long grouping between times, got together with putrescible waste, loose covers, and hurt compartments much of the time achieve inconvenience issues. An impressive part of the huge urban networks of the world depend upon a near to landfill to require their bit by bit squander; these zones are if all else fails from the start set up at the edges of the city. Objectives set up for landfills need to be sufficiently open and broad to suit the proportion of solid waste and other surface a city makes inside the course of 9-15 a long time. For expulsion in most clearing metropolitan landfills, rubbish is regardless taken to an exchange territory where it is washed down from the variety vehicle and stacked into a compactor or incinerator to diminish the volume. It is by then delivered to the landfill, which may be neighborhood or a long clear absent (Klotz, 2010).

Key to the suitable movement of exchange stations is the fast planning of deny. Despite their capacity, exchange stations routinely pull in flies, rodents, and inconvenience fowls, and their closeness can cause issues in wrapping neighborhoods. Free or compacted waste at the landfill is typically made sure about to reduce aroma and the intrigue it must different irritates. Right when soil is closed off, various materials, for instance, paper squash, parceled plastic, and sand, can be utilized. In mastermind sunshine sheets made from plastic make inside the focal deny a microclimate with temperatures tall satisfactory to imagine fly new development. Regardless, these sheets may lead to an interference with water penetration and common compaction, and trap landfill gasses. Crickets and cockroaches, checking the German cockroach, can ended up set up at landfills, dependent upon neighborhood conditions. Attacks of cockroaches have been related with secured bundles of nuclear family material that went to the landfill tormented. Once at the site, masses were kept up by the accessible food and simply confined compaction to smoothly harborage. The bothering feathered creature species changes consenting to zone; in any case, the primary customary are gulls, crows, starlings, and kites. They now and then settle at the region, nonetheless, by and large merge the landfill inside their looking through locale.

The gritty shaded rodent is ordinary in landfills globally. Gigantic vertebrates, for instance, foxes, wild pooches, and goats additionally routinely occur. There could be scarcely any predictable conditions unequivocally on the landfill to support vertebrate people groups; some bugs so to talk move to the landfill for taking care of and have created homes offsite. Regardless of the way that there's a steady wellspring of waste, the working face for dumping changes and there's common agitating impact through laborers also motor cars. Startling upsetting effect of cockroach, cricket and housefly masses can realize the dispersal of large populations to ranges joining the landfill. Blow flies and houseflies are capable of journeying up to 3km from swarmed objectives, and cockroaches can travel beyond a moved scene to building borders.

Colossal amounts of seagulls at landfills can agitate the movement of compaction and earthmoving contraption and spread illness. Waste at landfill areas from gulls have been appeared to contain human pathogenic microorganisms (Klotz, 2010).

Landfill gulls have the ability of transportation of microorganisms to create and urban objectives. Urbanization sway has enunciated the impacts on the abiotic parts of the earth. Centralizations of warmth holding surfaces of avenues, interstates, ending parts, the restricted proportions of open soil and green soil, and tremendous totals of tainting and particulate issue inside the discussion about result in urban zones having an air extraordinary comparable to the including open nation. Changes in the climate can happen inside the condition of ordinary temperature lows and highs, in elevated and heading of the breeze fields around structures, and in entire of precipitation and flood conditions. Air is the net blend of temperature, water seethe inside the air, precipitation, sun-based radiation, and speed of the breeze. Meteorological factors that are if all else fails certainly different among urban zones and open nation merge temperatures at night and day and relative clamminess, precipitation, plus haze.

8.5 IRRITATION CONTROL AND NUISANCE STATUS

Irritation status for arthropods in the urban condition is as a result, somewhat, of continued with proximity of a creature bunches in and around workplaces plus living areas. Adding to this is the potential clinical or mental reaction and money related hardship associated with their occasion. The continued with proximity of these animals is normal, somewhat, to the general insufficiency of control measures, and the nearness of flexibly regions and masses that oblige reinfestation. Long stretch assurance additionally, trouble status of private and peridomestic arthropods are considering an arrangement of little intrusions in by and large inconsistent living spaces, and huge store peoples in modestly stable normal environmental factors (Dhang, 2014).

Gracefully living spaces give individuals or social affairs that can revive neighborhood invasions and develop new ones. Without their flexibly masses, by far most of the ordinary vermin species would not bolster the abundance fundamental for trouble status. Vermin status is commonly associated with a certified or seen clinical risk, a tireless disturbance, or on financial mishap. The greater part of arthropods in this condition fit the bill for one or of course a more noteworthy measure of these classes.

Aggravation status could tamper and alter population of bugs. It could start by an aggravation through closeness of little amounts of bugs, by then become a prosperity risk by the proximity of colossal numbers, and over the long haul a money related level is shown up at when control and fix are required. Peridomestic pesters, for instance, umbrella wasps or yellow jackets settling in shades of homes or underground termites hurting essential wood, may acquaint a peril with human prosperity or mischief the physical structure of a structure. Family unit bugs hurt food, surface, and various materials, yet moreover jump in on near and dear space to cause mental pressure. Bothering status may be established solely on an upscale or eager reaction to the closeness of a frightening little creature or other arthropod, for instance, dreadful little creatures and centipedes. The money related and clinical purpose behind trouble status is quantifiable, anyway may be material to only a select social affair of bugs. Aggravation control exercises subject to snazzy excited reasons are generously less quantifiable and obvious, anyway are no less critical also undoubtedly a reason for countless control measures in the urban condition (Dhang, 2014).

8.5.1 Injury to their Aesthetic

The bothering status of specific arthropods in family and peridomestic situations relies only upon an extremism of their quality. For certain people there is a psychological or excited affectability to the closeness of a dreadful little creature or other arthropods. The living space is an individual and sacred recognize, the proximity of bugs or various animals may direct impact the individual fulfillment there, and their quality is commonly seen as prohibited. Obstruction for animals in this space is typically low, and control relies upon an energetic or elegant cutoff. Food spoiled with remote issue is prohibited on smart and general prosperity premise. Nevertheless, bugs, and various arthropods are so all inclusive accordingly different that couple of, accepting any, food can be freed from in any occasion a constrained amount of mischief or then again debasement by them. At the point when everything is said in done, government associations have set up most extraordinary degrees for standard or unavoidable deserts in sustenance for use by humans presenting no prosperity risk (Kalman & Crossingham, 2006).

Doubt the deformations, that are for most parts in the kind of live or dead frightening little creatures, body segments, and other common material, are harmless relies more upon comprehension than on attempt. It is typical that, if any risk to human prosperity were perceived to be connected with these admissible defects, the obstruction levels would be adjusted for human prosperity. The typical client may grasp and recognize that unadulterated food, with the ultimate objective that it is freed from all corrupting, may be problematic to achieve in an anticipated manner, anyway that over the top polluting by dreadful little animals or other material is unacceptable, at any rate on a tasteful premise.

A stylish physical issue is a decision limit for a bug control action that resembles the money related edge applied to agricultural bugs. The money related breaking point is a purposeful vermin thickness at which control moves should be made to shield an irritation masses from showing up at the budgetary injury level. In the urban condition, classy thoughts rather than money related ones are much of the time fundamental in beginning control exercises. Tasteful injury may be connected with a specific number of individual aggravations. Inside, many notable arthropods which caused a control action at a low thickness are cockroaches, silverfish, moth flies, plus floor covering frightening little creatures. Obstruction for incidental aggravations, for instance, ants, regular item flies, bunch flies, also parasite could be somewhat higher probably as a result of their standard occasion. Outside, aggregates of frightening little animals routinely lead to control exercises; essential bugs in this order are boxelder bug, ladybird scarab, elm leaf bug, and cricket.Large amounts of chironomids, winged ants, and mayflies may be a disturbance, anyway control measures are ordinarily not practical (Kalman & Crossingham, 2006).

8.5.2 Medical Effects

Many demands of frightening little animals as well as various arthropods have species which have clinical hugeness, considering the way that they snack, sting, suck blood, transmit parasites plus pathogens, or considering the way that they impel hypersensitivities, impulsive parasitosis, or entomophobia. No remedially noteworthy irritation has a solitary urban flow; all occur in urban and regular common environmental factors, to a progressively noticeable also lesser degree. Regardless, at times the bugs occur in and around working spaces and living areas, their centrality additions also control exercises are continuously typical. Arthropods with the most raised vermin status are those which convey a troublesome eat, sting, and suck blood. In spite of the way that they may present only a compelled prosperity peril, their quality isn't persevered. The most notable of these general fuse head bug, scabies parasite, blood suckers, and creepy crawlies (Kalman & JCrossingham, 2006).

Bites, Stings and Blood Sucking Insects: Leeches, scabies, and lice happen regularly among mankind masses, plus at all budgetary levels far and wide. People differentiate in their reaction to these arthropods: some are insignificant affected, anyway in case dealing with continues and masses shoots up honing happens. The abundance of scabies and lice radiates an impression of being cyclic in a number of industrialized countries, anyway logically ordinary and less cyclic in making countries. They are typically found on fundamental school children, and there is oftentimes a social disfavor related to embodiment of these bugs.

Figure 8.1: A picture of a bloodsucking bedbug which leaves marks on the human skin.

Source: https://theconversation.com/in-defence-of-the-bed-bug-54218

Lice plus scabies are moreover ordinary in wartimes also starvation at times that huge amounts of dislodged individuals, poor sterile conditions, plus amassed everyday situations. Leeches are relatively associated with individuals. These blood-dealing with parasites are dispersed far and wide, and irregularly they become different and intrusions increase in private and business structures. Positive conditions present indoors, snappy improvement of people and materials globally, and lessened bug splash use inside could have added to the return of these domiciliary aggravations. Despite the physiological response or conditions, people scorn these ectoparasites because of their substance, as well as their effects on individual fulfillment. Vermin status of lice, scabies, also kissing bugs could be established on the ugly condition of the polluted skin plus shivering also disquiet achieved by their dealing with. Irritation status of dreadful little animals is in a general sense classy since the vast majority of those found inside are not inclined to eat or be a prosperity threat. There are two or three creature types that have a troublesome snack, once in a while with extraordinary outcomes. Practically all bugs are poisonous, in any occasion with regards to their ordinary prey, anyway just around 20 of the about 30,000 depicted types are perilously hurtful for individuals. The most critical species are: the commanding house bothers that oftentimes eats people without actuation; yellow sac 8-legged creature that are found globally plus kinds of recluse also widow bugs. The snack of these 8-legged creature is normally troublesome plus venom could be locally and on a very basic level unsafe (Rufer, 2015).

Transmission of parasites and pathogens: Mosquitoes, ticks and reduviids transmit the huge arthropod-borne ailments in the urban condition. Majority of the vectors happen essentially in nearby peridomestic regular environmental factors, or immediately move to these living spaces from store masses outside urban zones. Their thriving and by and large dissemination are arranged incompletely on their ability to change in accordance with new has or substitute their trademark replicating areas for those available in or around human homes. In general dispersal of the mosquito species are associated with its change in accordance with human characteristic environmental factors, for instance, its ability to bring up in counterfeit compartments and to go with individuals. Around homes are various compartments containing water and viably substitute for the innate tree opening conditions of this species. The limit of the eggs to suffer drying obliges critical separation transport to new locales. This species bears best where there is openwater accumulating and counterfeit compartments (Robinson, 2005).

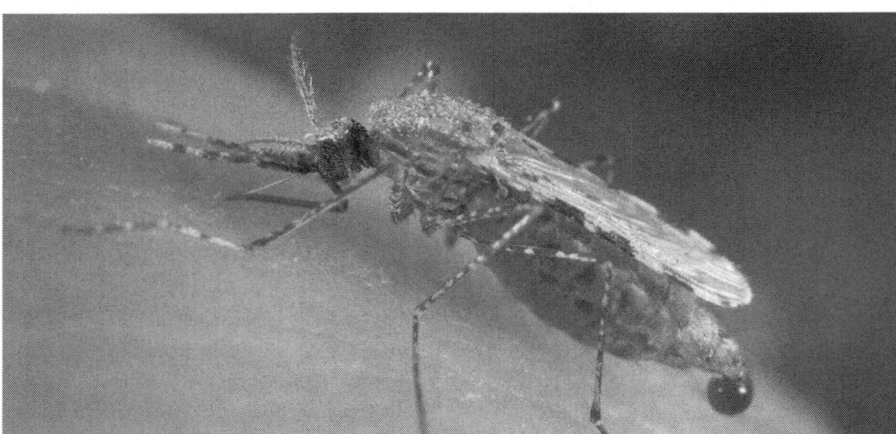

Figure 8.2: A picture of a mosquito, a blood sucking insect that causes malaria which is one of the deadliest diseases in the world.

Source: http://www.ox.ac.uk/news/2016-06-27-genetic-study-shows-how-malaria-mutations-beat-treatments

Due to its alteration limits plus occasion in urban circumstances globally, it has a capacity of transmitting new arboviruses which could frame as common pandemics. Another species is *Ae. Albopictus*, that has subbed its trademark recreating site of plant pits for nuclear family compartments plus vehicle tires in the urban condition. It has broadened its transport from Asia, presently it is a huge disturbance in urban also country zones, plus a noteworthy dengue vector in urban areas. *Culex tarsalis*, the vector for western equine encephalitis contamination, CX.tritaeniorhynchus, the vector of Japanese encephalitis contamination, are nation species yet enter urban normal environmental factors after generous precipitation and flooding. Their suitable breeding grounds are those that hold water. Cx. pipiensis a common notable disturbance type in urban conditions, plus it transmits a couple arboviruses. Subspecies Cx. pipiens quinquefasciatus is the huge mosquito vector in urban circumstances all through the world. It breeds in ground pools and in water that accumulates in nuclear family compartments, and expeditiously goes into houses.

This species is all around changed in accordance with urban and current conditions, and it is an overall creature types in the septic outskirts in making countries. In India, An. culicifacies normally breeds in like manner waters, yet will duplicate in overpowered burrow pits and pools in urban zones. Flooding plus overpowering precipitation give raising goals to An. atroparvus, An. messae, An. sacharovi, and An. superpictus in urban regions in Europe. The flight extent of these species, and their dispersal by the dominating breeze at the edge of urban zones, has influenced the urban occasion of these and other anopheline mosquitoes (Robinson, 2005).

Peoples of An. gambiae occur in urban domains in sub-Saharan Africa, where it breeds in underground stockpiles and catch bowls of whirlwind drains in urban zones. Trypanosoma cruzi is kept up in the urban condition in the family unit and peridomestic peoples of cats, dogs, opossums, armadillos, squirrels, and a couple of sorts of rodents and mice. The basic vectors are Triatoma spp., Rhodnius prolixus, and Panstrongylus megistus. The point of convergence of this disease is the poor family conditions in provincial regions and septic edge of urban networks. Dermacentor, Rhipicephalus, and Ixodes ticks find proper conditions and has in the greenspace and peridomestic situations. In rustic districts there are many people who have been infected with Unpleasant Mountain spotted fever (RMSF) plus occurrences of Lyme ailment in USA also various countries. An extension in RMSF is connected with the accomplishment of Dermacentor variabilis in rustic vegetation and possibly wealth of family unit hosts such as dogs and others. The extended recurrence of tick loss of movement follows the abundance of D. variabilis in provincial zones e.g. the eastern parts of USA also high population of Ixodes holocyclus in Australia (Robinson, 2005).

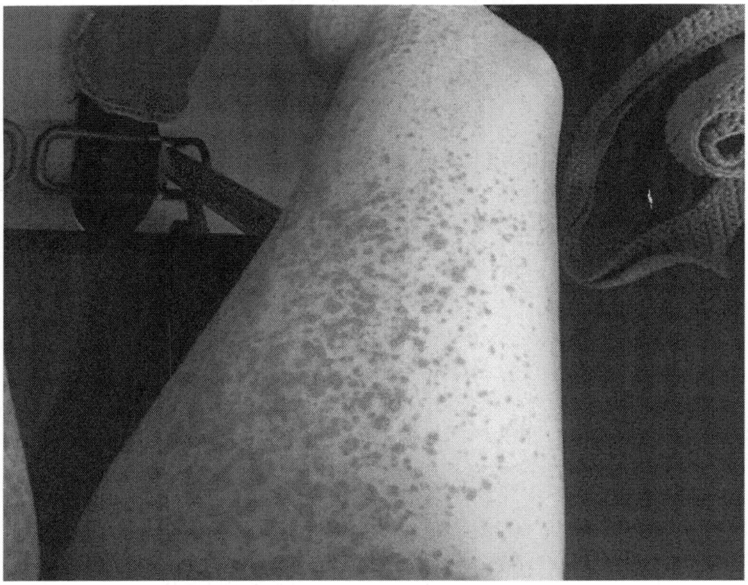

Figure 8.3: Rough Mountain spotted fever and how it appears on the human body.

Source: https://www.wrcbtv.com/story/36646454/what-is-rocky-mountain-spotted-fever

In the Sydney suburbs, I. holocyclus is plenteous considering the way that the mix of standard vegetation in peridomestic living spaces gives fitting conditions to the bandicoot the tick's host. Lyme contamination is one of the most notable arthropod-borne diseases in provincial districts the world over. I. scapularis is the focal vector in northeastern USA. The energetic periods of this tick feed on different winged creatures, warm blooded animals, and individuals; the whitefooted mouse is the fundamental store and vector to individuals, and the white-followed deer is the basic site they overwinter and host for the adult tick. Adults don't move from host to have and don't transmit the affliction.

The dispersal extent of scapularis is developing in country zones, close by the event of Lyme ailment, on account of the extension of deer in these conditions. Rhipicephalus sanguineus completes its progression on private and non trained pooches, also it has acclimated to urban conditions in various bits of the world. This species started in Africa, and has spread to America, Europe, Asia, as well as Australia, where it is settled. This tick needs decently high temperatures to complete improvement. In gentle nations globally it is related to dogs inside; in warm environments happening outside in provincial zones and is a vector of rough mountain seen fever in the USA and a vector of boutonneuse fever of the Mediterranean district in Europe (Robinson, 2005).

8.5.3 Hypersensitivities/ Allergies

Negatively helpless disease is an ordinary issue impacting an estimated 40% of all out masses. The allergen proteins which prompt negatively defenseless reactions could be taken in, ingested, also held on the skin, and mucous layers. Typical easily affected reactions Fuse developing, disturbed and watery eyes and nose, inconvenience respiration, cerebral torments, rashes on the skin, also shivering. Various kinds of arthropods are the wellsprings of allergens that hone and cause horribly defenseless reactions in individuals.

These allergens are proteins also physiological response to introduction is comparable to it is for separate allergen sources, for instance, plant residue, molds, and a couple of sustenances. Arthropods in the urban condition which incite horribly helpless reactions in individuals join flies, bugs, bugs, and moths in set aside food, and stinging dreadful little animals, for instance, ants, wasps and bumble bees. In any case, the high number of lurking cockroaches and buildup vermin in the living spaces and their ground-breaking allergens make these two critical wellsprings of excessively touchy reactions. Cockroaches customary in also around homes are a critical wellspring of allergenic proteins.

The number of people affected by cockroaches is high globally, ranging from 23 to 60% of the people; it is obvious as respiratory asthma and dermatitis. In some midtown neighborhoods in the United States of America, an approximate 37% of children could be defenseless to cockroach allergens. The cockroaches known to be the wellsprings of allergen fuse kinds of Blattella, Blatta, and Periplaneta. Regardless, Blattella germanica and P. Yankee fables are the transcendent indoor aggravations, and contribute the most to clinical issues. The allergens from these frightening little creatures are found in the fecal material, oral releases, exoskeleton pieces, and dead bodies. Particles bearing cockroach allergen are basically carried on particles under 10 μm separation over; these particles settle quickly and reduce presentation (Robinson, 2005).

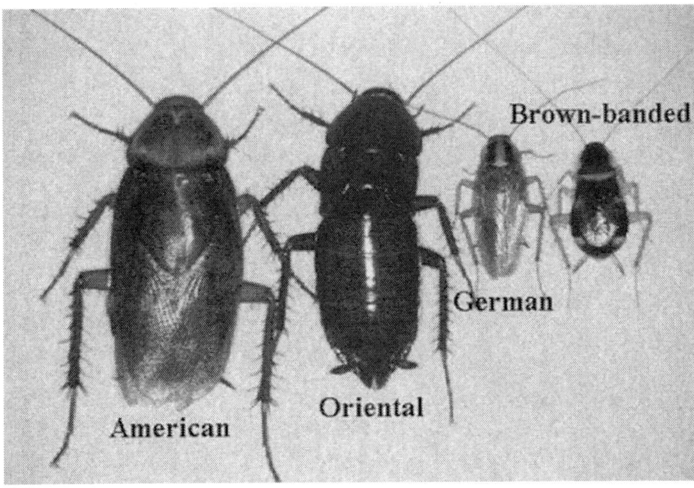

Figure 8.4: This picture represents the different blattella, blatta, and periplaneta cockroaches.

Source:http://www.dbmonline.net/globalpest/content/nm/extra/

Buildup vermin are in taken care of food things and have the living and working space far and wide, and they are the wellspring of allergens. Affectability to bug allergens is eminent; in the USA and Europe 20–35% of horribly helpless individuals are sensitive to clean bugs. Most homes and work environments involved by dust vermin contain a couple of creature types, including Dermatophagoides farinae, D. pteronyssinus, D. microceras, and Euroglyphus maynei. Buildup vermin masses require a wellspring of protein-rich food and characteristic conditions of 10–30 ◦C and in any occasion half relative tenacity (RH). The 0.5–1.0 g of skin scales individuals shed every day give satisfactory food, and floor covers and material materials on beds, furniture, and clothing give harborage and duplicating goals to these vermin.

Allergens of Dermatophagoides species are made in the back midgut and hindgut as stomach related synthetics, released fecal pellets and in cast skins of bugs. These allergenic particles are commonly enormous also quickly fall in undisturbed air. In any case, dung pellets growing dry also segment, and little particles may get airborne. Parasites related with set aside sustenances also developments join Acaris siro, Glycyphagus domesticus,

Lepidoglyphus destructor, and Tyrophagus putrescentive. They feed on structure and developments that create on nuclear family sustenances, however then again are found on materials and on divider and rooftop surfaces (Kalman, 2009).

Most species need about 70% to 98% tenacity for headway. Prologue to take care of food vermin allergens may be by ingestion or through internal breath, also refinement to these bugs has been represented in many made countries. The jumbled flour bug, Tribolium confusum, is apparently the most notable contaminant of flour, grain, orchestrated flour mixes, dried nourishments developed starting from the earliest stage, various flavors. In the provided materials, there are chances of areas with all the presence periods of this bug, and for some attacked material, there may be live adults and hatchlings. Individuals that are unstable to frightening little animal allergens could have a horribly defenseless response when ingesting spoiled flour things.

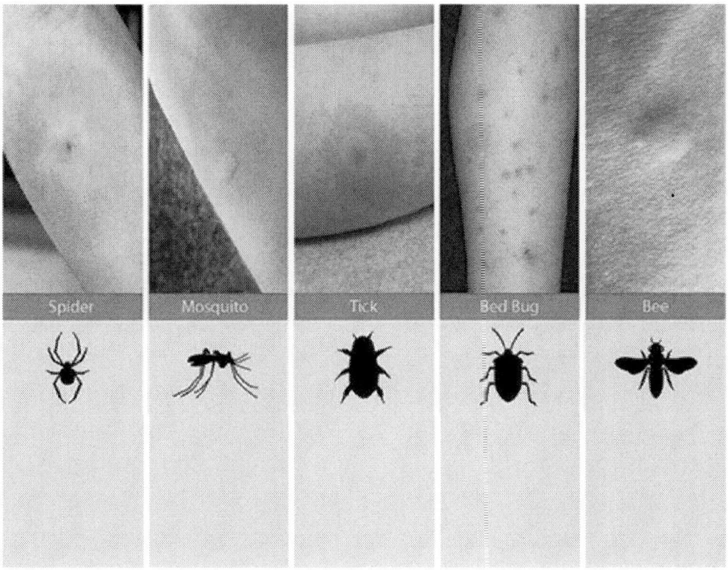

Figure 8.5: This picture shows different skin allergies caused by insects. The insects indicated are responsible.

Source: https://www.toppr.com/content/story/amp/entry-routes-to-allergens-44470/

8.6 METHODOLOGIES TO CONTROL URBAN PESTS

Control of family and assistant aggravations joins the useage of an array of engineered plus nonchemical strategies. General size of urban and country zones plus the noteworthiness of private also peridomestic annoys give the budgetary spark to producers to figure bug splashes and design control programs for essential bothering species. Thought for disagreeable prologue to nontarget species and individuals has improved the movement of current bug showers, while keeping up or extending suitability. Incorporated bug the board (IPM) began in the agroecosystem that has the necessity to give a money related base to decisions to use creation control. The objective of IPM was to give a convincing and monetarily gainful approach to manage trouble control. The forefront thought of aggravation the administrators is the compromise of natural, substance, and other control procedures into a program that limits bug thickness to levels underneath those causing money related injury. The critical component of cultivating IPM is covering of masses and not obliteration of vermin peoples. Vermin thickness is associated with a money related physical issue level, which is seen as the most decreased masses thickness that will cause fiscal mischief. Damage could happen underneath this level, yet it is seen as commendable since it doesn't impact yield or the estimation of the product. Lessening pesticides use and directing aggravation masses has been a consideration for the urban natural framework (Kalman, 2009).

Within the peridomestic condition, pesticides are basically used considering ordinary vermin that cause a la mode hurt. Activities that keep up these annoyances at low levels may be fitting, organically significant, and financially beneficial. In any case, applying the standard IPM hypothesis of vermin the board to arthropods in the nearby condition could be irksome or not acceptable. Control exercises that have the main purpose of keeping up bug masses underneath a money related or up-to-date level may not be necessary for frightening little animals that are a clinical peril to people or pets or that may cause helper hurt. A couple of vermin in the family unit condition probably won't have a fiscal or prosperity related level of aggravation status; regardless, their substance in the home gives little space to strength and satisfactory explanation behind their end. There is minimal confirmation that the possibility of a level of obstruction can be applied to nuclear family disturbs. Urban occupants normally adjust their mindset toward the proximity of irritations in the home, the level of their vermin strength rots as levels of infiltration lessens (Kalman, 2009).

Life forms debase for all intents and purposes all dreadful little animals and various arthropods. Entomophthorales fuses a couple of creature types that are lethal to soilinhabiting dreadful little animals and termites. Soil conditions give a consistent area to these living creatures; regardless, the upper 2–5cm of various soils show up at temperatures that are destructive to the vegetative periods of most pathogenic living beings. Metarhizium, Beauveria, also Verticillium have species that are pathogenic to frightening little creatures. The most regularly inquired about species consolidate M. anisopliae, B. bassiana, plus V. lecanii, all of which have a wide host go. Strains of M. anisopliae have been used to control underground termites and indoor cockroach annoys. Sorts of the development assortment Cordyce pioideus recorded termite settlements present in countries such as Mexico as well as Kenya (Collins, 2012).

8.6.2 Nematodes

Nematode species in the families Steinernematidae(Steinernemas pp., Neosteinernema spp.), Heterorhabditidae (Heterorhabditisspp.), and Mermithidae parasitize frightening little animals. Hatchlings of Steinernematidae pass on the pathogenic organisms, Xenorhabdus nematophilusand X. luminescens. Photorhabdus microorganisms are passed on by roundworms in the family Heterorhabditidae. These minuscule living beings can enter the dreadful little creature have and execute it inside a period between 24hrs and 48hrs. Nematode hatchlings are free swimming in water also pollute the bug have entering through the mouth, backside, or spiracles. In the sorrow of the body, the hatchlings release the minuscule life forms. The nematode profits by the microorganisms cells plus the separating tissues. Right when substrate is drained, they leave the dead body plus search for another host. Nematodes are normally secured also satisfactory, plus they are applied easily along standard sprinkle spouts.

Restrictions of the control methods of nematode bacterium are that free water is key for their host-pursuing behavior, and these animals have cause difficulty to bring up in mass social orders. The nematode Deladenus siricidicola cleans the wood wasp, Sirix noctilio, and has viably been used in the decrease wood wasp masses in pine tree farms in Australia.

8.6.3 Organic Bug Sprays

Plant things for dreadful little creature control have been taken up as attractants, enemies of operators, as solvents, and transporters of bug splashes. In any case, the basic usage of plant blends is as toxicants. Nicotine, expelled from the plants N. rustica and N. tabacum, has been used for quite a while to slaughter dreadful little creatures. The technique for movement is to impact the central tactile arrangement of the target animal genuinely. Limonene is expelled from citrus strips. It is fruitful against some outside parasites, for instance, bugs, lice, vermin, and ticks, it is nontoxic to warm blooded animals. The common impacts the material nerves of the periphery tactile framework; is anything but a cholinesterase inhibitor.

These engineered mixes are isolated from blooms of a couple of creature types, including Tanacetum cinerariifolium, T. carneum, and T. coccineum. This strategy for movement is to attack the periphery tactile framework, furthermore, the brief effect is a quick enthusiasm also knockdown. At low estimations some recovery can occur, due to assimilation of the pyrethrins. They are ordinarily nitty gritty with synergists, for instance, piperonyl butoxide, to decrease the oxidation technique plus augmentation sufficiency. The social affair of bug splashes is nontoxic to warm blooded animals.

8.6.4 Inorganic Bug Sprays

Inorganic bug showers have a long history of use in nuclear family plus set aside edible trouble control and this get together consolidates arsenic, boron, mercury, and sulfur. They are consistent blends and noxious to a far reaching extent of animals. Sulfur is mostly settled to pesticides, plus is toxic as a contact or stomach poisonous substance to bugs, also scale frightening little creatures, plus as a stomach poison for a couple of caterpillars. Sulfur cleans plus its showers are furthermore fungicidal.

8.6.5 Natural Bug Sprays

A large portion of normal bug splashes apply a harmful effect on segments of the dreadful little creature tangible framework. The tactile arrangement of frightening little creatures and very much developed animals is reliant on the transmission of nerve inspirations along the axon, all sent from the cell body, across transitional synapses to the nerve polishing off with the muscle. At the nerve completing a transmitter substance, gamma-aminobutyric destructive, is released, which realizes muscle withdrawal. Main thrusts come the axon by virtue of changes in the electrical potential, including potassium and sodium particles, over the outer layer of the axon. Main thrusts travel along the axon and definitely show up at an opening between two nerve endings, the neural association. Transmission of an inspiration over the synapse is intervened by the manufactured acetylcholine, which is released at the outside of the axon film. Acetylcholine moves over the gap, is gotten by receptors on the contrary side, besides, another inspiration is then made in the confining axon. To thwart assortment of acetylcholine in the synapse, acetylcholinesterase is released. It is isolated to choline and acidic destructive. A couple of the consistently used bug showers square production of the esterase. This thwarts the section of dynamic messages in the nerve, and this may incite breakdown of the on edge system and end.

8.6.6 Anti-specialists Aka Repellants

Enemies of specialists shield a disturbance from showing up at a edible source habitat, or relocate it once there. They are generally referred to as nontoxic to bugs, plus nontoxic, nonirritating, in addition, nonallergenic to individuals and neighborhood animals. A huge part of the earlier enemies of operators substances had strong, noticeable aromas the progressed, built enemies of specialists are nearly unscented. Despite the way that rebuffing aggravation dreadful little animals, for instance, body lice, biting flies, and floor covering frightening little creatures has a long history, the front line utilization of this thought abuses safe use and application features, and typically negligible exertion. Protection is ordinarily present second, and may be suitable for not many species.

8.6.7 Attractants

The material instrument related with searching for edibles, oviposition goals, also mates is vivified and constrained by manufactured inventions. Manufactured substances that pass on lead messages or activate a response are named semiochemicals, plus pheromones are semiochemicals that are used in intraspecific correspondence between individuals of comparative species. There are a couple of novel sorts of pheromones, for instance, caution, dispersal, and amassing pheromone, anyway it is the social affair of sex pheromones that is most comprehensively taken up in bug the board programs. The central jobs of pheromones in control and the board programs are: male getting, which diminishes the regenerative capacity of an irritation people; mating interference, which upsets the mating search of folks; and in the acknowledgment likewise, checking of adults to choose masses abundance additionally, zone of attacked material.

8.6.8 Insecticide Gases

Methyl bromide is a fumigant bug splash which rapidly butchers bugs, parasites, and nematodes. It enters substrates including soil and wood, it customarily doesn't recolor or degenerate things, and is noncorrosive and nonflammable. It has a foaming reason for 3.6 ◦C and is horrid and unscented at obsessions used for fumigation. Chloropicrin is from time to time included at 2% as awarning marker when this methyl bromide is used in structures. Strategy for action is mischief to nerve cell films, also, it reacts with sulflhydryl bundles in proteins. Frightening little animals for the most part fail horrendously inside 24 h of presentation, anyway mortality could be conceded about 48hrs.

9

Stored Products as Habitats for Insects

CONTENTS

9.1 INTRODUCTION

The term 'stored products as habitats for pests' can be used to refer to any creature that damages after infesting carpets, leather, fabrics, books and documents, stored food, and any other preserved or dried item that is not utilized immediately after it is has arrived at location or regularly moved. Generally, these organisms can be inclusive of vertebrates such as birds and rodents, arthropods such as mice and insects, and microorganisms such as bacteria and fungi. These tend to be responsible $4 million worth of lost money each year in products that have been contaminated, and even in the destruction of important heritage artifacts and documents in museums, offices and homes. Majority of these organisms end up in the buildings through items that were initially infested when they were being bought. Others end up indoors when there is poor storage of items that are susceptible to them, or when single pests find a way to access them. Majority of the time, most of these pests are not noticed because they tend to infest products that are hardly ever use and many of them are actually quite small in size.

Figure 9.1: When insects use stored products as a habitat, they tend to destroy the products, using the products as their source of food. If not handled, it can become an issue of economic importance.

Source: https://bpca.org.uk/News-and-Blog/spi-control-how-to-get-rid-of-and-prevent-stored-product-insects-bpca-a-z-of-pests

We tend to notice infestation when these pests emerge from the stored products, maybe due to crowding, to disperse, or after they have exhausted the specific stored product and are now searching for new products that they can infest. Other indicators of possible infestation by storage pests may include unexplained occurrences of beetles and small moths crawling over ceilings, walls and counter tops or flying around stored items in large numbers, stale odors in closets and pantries, and powdery residues surrounding and below such stored items (Hagstrum & Subramanyam, 2016).

It is also possible to detect infestation in stored beans or whole-grain after soaking them in water, and seeds that are hollowed out float up to the surface, together with the pests' adult stage and other debris. Other signs include webbing or clumping of particles. It is very important to find these organisms and also find out the source of infestation immediately, so as to make sure that there is no more establishment and they do not also spread.

In this specific write up, we will be discussing mites and insects that are commonly found in stored product and we can group this into stored fabric pests and stored food pests. It is also important to note majority of these organisms have no restrictions to one specific group and can also be found in fabric, food, as well as other items similar to this (Hagstrum & Subramanyam, 2016).

9.2 STORED FOOD PESTS

The different insects that we will talk about in this section are grouped into non-insects such as mites, and insects such as beetles and moths. The Beatles and moth larvae are all cream or pale-white and they are normally hard to distinguish; however, the adults can be identified quite easily. Majority of the adult beetles are normally located together with the grubs as they feed on the stored products; however, the adult moths never feed on these products but they use these products to lay the eggs. It is not possible to identify mites with the naked eye and they require magnification to be seen.

In general, when these pests are eaten in small quantities together with the food products, they do not cause any harm. However, when these products are largely infested, the quality, flavor, taste, and appearance of the food may be significantly altered, as well as lead to the creation of conditions that are favorable for the growth of secondary bacterial and fungal pathogens. It is also possible for sensitive people to suffer from allergic reactions after ingesting these products. Other pests such as ants may be found in the stored foods, and these can be harmful if consumed in large quantities (Hagstrum & Subramanyam, 2016).

9.3 BEETLES

9.3.1 Pulse Beetle, Seed Beetle, Bean Weevil

Above are the common names used to refer to *Callosobrunchus* spp, and *Acanthosceldes* spp. These species are in the order Coleoptera and the family Chrysomelidae. They can be found on every single part of the world. These beetles are quite small, being approximately an eighth to a quarter inch in length. Their bodies are stout, rounded and oblong and they come in either dark gray or grayish -brown with mottled patterns. Even though they are referred to as weevils, they do not have the characteristic elongated snout that true weevils have.

Figure 9.2: It is possible to unknowingly mistake infested beans for those that are healthy, because when bean weevils are present, is not easy to notice them until you find empty husks after the crops have been harvested.

Source: https://www.orkin.com/other/weevils/bean-weevil

Their larvae are yellowish cream in color and are small and plump without legs. The best way to identify these weevils through their short elytra which refers to their front wings which are hardened and form a shell on top of their body. This short elytrum has flattened ends that do not reach the abdomen's tip. Their heads are small, bent down words, and black. They have large and protruding eyes and their whole body is covered in fine short hairs. They have well-developed antennae, and most of the time they are mostly as long or even longer than their front legs. Their antenna also has certain segments that are flattened and leaf-like towards the tip. When the infested beans are disturbed, the adult beetles are able to fly out in large numbers (Hagstrum, Phillips, & Cuperus, 2012).

These pests are well known for the damage that they inflict on stored legumes such as beans and peas, and sometimes they also destroy grains and other foods. They also have the ability to attack legume pods that are still on the field. They normally infest the seeds by boring into them and feeding on the contents inside them. The damaged seeds have neat and circular exit holes. When the pests heavily infest these products, the beans can end up heating up and this could lead to the growth of mold and even the alteration of taste and flavor. These weevils are also capable of chewing through paper bags and plastic to access the stored food.

When it comes to their life history, they normally lay their eggs on the surfaces of the beans where they stick and can be seen as small white spots. There may be several eggs on the surface of each seed. These weevils are also able to lay the eggs on the beans that are still in the field before they are harvested. When the larvae emerge, they bore into the seeds and feed on the content inside the seeds to feed on many seeds before they are able to pupate. When they are ready to undergo pupation, they normally chew a circular hole on the seed where they can come out as adults, and then they leave the seed after pupating inside it. The pupae are normally dark brown. They are small. They are highly dependent on temperature for the duration of their lifecycle and it takes approximately 3 to 4 weeks at temperatures above 30°C, and this period can be much longer in cooler months. In one year, they can have approximately 5 to 6 different generations.

9.3.2 Rice Weevil

Above is the common name used to refer to *Sitophilis oryzae*. This species is from the order Coleoptera and family Curculionidae. This species is commonly mistaken with two different species that are closely related: the granary weevil or wheat weevil (*Sitophilus zeamais*) and the greater rice weevil or the maize weevil (*Sitophilus zeamais*). These weevils can be found in all parts of the world.

Figure 9.3: Controlling rice weevils requires for one to locate and remove all food sources that are potentially infested. In all stages of development for the Rice weevils one can kill them by freezing food that is infested below 0 °F (-18 °C) for three days.

Source: https://www.slideshare.net/SangeetaBhalerao2/rice-weevil-78741141

The weevils are normally either black, reddish- brown, or dark brown in color. Their bodies are elongated and are approximately a 1/16th an inch in length. They have plump, small, yellowish- cream larvae. The best way to identify them is through their characteristics snout that projects at the front of the head, having antennae that are elbowed and have tips that are club -shaped. When observed with a microscope, it is possible to see four orange or reddish brown patches on the corners of the wings, especially in maize weevils and rice weevils, with those of the maize weevils being lighter. Both of these weevils have the ability to fly. These marks are not found on the granary weevils, and it is also not able to fly. For all the species, when they are disturbed, the adults normally go motionless, with their legs withdrawn closely to the bodies and pretend to be dead. The larvae in all the species are legless, plump cream-colored, C -shaped and small (Hagstrum, Phillips, & Cuperus, 2012).

These pests are very important as stored food pests of grains, mainly corn, wheat, rice and all related products such as noodles, macaroni, breakfast cereal, etc. The adult maize and rice weevils also have the ability to damage grain crops that are still standing in the field before they are harvested. All the species are also capable of attacking other different stored foods such as nuts and dried fruits.

The adults chew on the grain and eat up all of the content, leaving just the outer covering, and thereafter, the larvae undergo development within these hulls. If they infest these grains for a long period of time, they can end up reducing it to a stale, coarse, smelly powder. When you look at the individual grain that has been infested, you can notice irregular feeding marks and some of these grains have only been hollowed out partially. When these greens are heavily invested, they heat up, and this can also lead to the growth of mold and the alteration of taste and flavor. These insects are also able to chew through paper bags in plastic to access the stored food.

When it comes to the lifecycle, the females normally chew small holes on the grain's surface, lay one egg inside it in then seal the hole with a waxy secretion. When the grub hatches, it feeds on the insides the grain into it pupates, and comes out of the grain as an adult through having to break through the remaining covering. These are also highly dependent on the temperature for the duration of the lifecycle and they also take approximately 3 to 4 weeks at temperatures that are higher than 30°C. They may also end up taking longer in cooler months. But these insects can end up having approximately ten different generations per year (Hagstrum, Phillips, & Cuperus, 2012).

9.3.3 Cigarette Beetle

The above name is the common name used to refer to *Lasioderma serricorne*. This species is from the order Coleoptera and family Anobiidae. This beetle can also be found in all parts of the world. It is brown, stout, reddish brown, and small. It is approximately 1/16th of an inch in length and its outline is oval.

It is best identified by the its thorax and head bent downwards almost at a 90-degree angle to its body, so much that it is not possible to see its head when you view it from the top. It has an almost smooth elytrum, and if not smooth it may have very faint grooves. It has fine short hairs all over its body and its antennae are short having a thickness that is uniform all through them. They produce small, legless, creamy- white, plump larvae with heads that are small and yellowish brown, and also have yellowish- brown silky hairs covering their bodies.

Figure 9.4: Cigarette beetles can be found in many different places. They have a geographical range that is wide, but mostly thrive in subtropical and tropical areas. Their activity has mostly been observed near areas where there is storage of grains or other food.

Source: https://www.rottler.com/pests/profile/cigarette-beetle

Most of the time, these beetles are confused with drugstore beetles, mainly because they are from the same family and tend to infest the same stored products. This is not the case however, because drugstore beetles have deep lengthwise grooves on the elytra, they have a three segmented club appearance on the end of the antennae, and their larvae have hairs that are much shorter than the cigarette beetle. When you put the two of them side to side and view them together, the drugstore beetles tend to look more elongated and slightly larger, while the cigarette beetles are more rounded and shorter. There is also another common confusion between these two beetles with another common beetles that is from the family Anobiidae, *Anobium punctatum* (the common furniture beetle), which tends to be longer than these other two, and on their pronotum, have an upward projection or pronounced hump that looks like a hood, and are also much darker in color.

These pests are important as they are stored -product pests for stored tobacco as well as other spices and dried herbs. It is also possible to find them in others stored products such as dried flowers, bonemeal, flour, oil cakes, cereal, and dried nuts and fruit. Cigarettes beetles normally prefer tobacco as their main food, and that is where they get their common name from. They attack all- processed, dried, and harvested forms of tobacco, including raw dry bales, chewing tobacco, cigarette, and refined cigarettes.

The grubs and adult beetles normally bore through the tobacco leaves which leaves holes and in the end can reduce the tobacco into a fine powder. They are also known to cause distraction to many other stored products in the same fashion (Hagstrum, 2016).

When it comes to their lifecycle, they normally loosely lay their eggs on the product, and the grubs that are hatching, together with that adults, feed together on the leaves until the product is exhausted. They also pupate on the same food products. When the larvae are mature, they have the ability to create small cocoons with their saliva and particles of the substrate, which they use to pupate. They are also highly dependent on temperatures for the duration of the lifecycle taking approximately 3 to 4 weeks with temperatures higher than 30°C. This can also be much longer in months that are cooler. In a year, they may have approximately 5 to 6 different generations.

9.3.3 Biscuit Beetle, Bread Beetle, Drugstore Beetle

Above is the common name for species *Stegobium paniceum*. This species is from the order Coleoptera and family Anobiidae. The beetle can be found on all the parts of the world. These beetles are elongated, reddish -brown and small with about an 8th to a 16th inch in length having outlines that an elongated and oval.

Figure 9.5: The major problem with these beetles is the level of contamination of the products they have infested. These beetles do the majority of their damage in their larval stage living, eating, and excreting waste inside the products they have infested.

Source: https://www.orkin.com/other/beetles/drugstore-beetle

They can best be identified with their thorax and head that are bent downward at almost a 90° angle to the body, to point where the heads cannot be seen when viewed from the top. They have deep and distinct lengthwise grooves on their Elytra. They also have fine short hairs all over their body and their antennae are quite short and have a three-segmented club on the tips. They produce legless, creamy white, plump and small larvae all which are also covered with silky hairs that are tiny.

As mentioned above it is actually quite easy to confuse the drugstore beetle with the cigarette beetle, mainly because they come from the same family and they tend to feed on the same store products. However, the elytra on the cigarette beetle are smooth and may have very faint grooves. The cigarette beetles also have antennae that have uniform thickness.

This pest is important as a pest of certain medicinal plants and dry herbs, in addition to other products such as leather, book bindings, cereal, dried fruits and nuts, tobacco, and dried herbs and spices. They tend to prefer medicinal plants and other dry herbs and they are commonly found in preparations for prescription drugs, thus giving them their common name. It is important to understand that they tend to also attack many other different stored products and have even been seen to bore through lead, in and aluminum sheets. These pests have also been seen to infest rodent and bird nests buildings and homes. Grubs and adult beetles normally bore through stored products and then leave tiny holes, and in the end this stored product is reduced into a fine powder. It is also not uncommon to find adult beetles chewing small shot holes on stored product packaging to get out, which causes some of the powder remains from inside the stored product to pour out (Hagstrum, 2016).

When it comes to their lifecycle, they normally lay their eggs loosely on the product and both the adults and the hatching grubs feed together on the product until it is exhausted. They also tend to pupate on the product. Just like with the previous insect, when the larvae mature, they are able to create small cocoons with the saliva and particles of the substrate which they use to pupate. They are highly dependent on temperatures for the duration of the lifecycle, taking approximately 3 to 4 weeks at temperatures that are higher than 30°C, and this could be longer in months that a cooler. In a year, they may create 5 to 6 different generations (Hagstrum, 2016).

9.3.4 Cabinet Beetle, Khapra Beetle

Above is the common name for *Trogoderma granarium*. This species is from the order Coleoptera and family Dermestidae. It can be found on all the parts of the world. This beetle is rounded and oval- shaped, stout, brown, dark and small with a length of approximately 1/ 16th of an inch.

When it is observed closely, it can easily be identified with the short dense hair that covers its body. It also has indistinct lighter brown or reddish brown patterns that may be found on the elytra. When viewed from the top, it may be possible to slightly view its head which is bent downward and the pronotum almost fully covers it. It has short antennae that end in a 3 to 5 -segmented club. They produce a quarter- inch long larvae which have bodies that are distinctly segmented with short, stiff hairs and a yellowish white color. They have longer hair towards the end of the tail. The larvae normally turn reddish brown or golden as they mature.

This pest is highly important, as it is considered one of the world's most invasive and destractive stored product pest, and it has even been placed on the quarantine list in many different countries inclusive of the United States. When this pest is noted, one needs to report this occurrence to the state regulatory agencies.

Khapra beetles normally invade and attack a high quantity of stored product of animal and plant origin, however, they have a preference to cereal products and grain products such as noodles, flour, whole grains, and other similar items. The larvae have been seen to cause more damage than the adults. When it comes to grains that have been stored, they are known to partially feed on a single grain before they start on the other one, and therefore one beetle can destroy a large amount of the grain. The adult beetles have the ability to chew through paper and plastic packaging to access the source of food. When they infest heavily, the product may end up heating up and this may lead to the growth of mold and alteration in the quality and flavor of the food items (Hagstrum, 2016).

Figure 9.6: The main reasons why this beetle is considered a nuisance include: its ability to survive at any location is storage facilities that have protection from environments that are cold, and its poor eating behaviors that include eating very little from one grain. Meaning that one beetle can cause a lot of damage.

Source: https://simple.wikipedia.org/wiki/Khapra_beetle

When it comes to their lifecycle, the adult beetles normally feed very little within their lifespan of approximately 15 to 30 days (which is fairly short), and they also rarely ever fly. They lay their eggs in a lose manner on the source of food and immediately after the larvae hatch, they begin to feed. Before they turn into adults, they normally pass through several molting stages. They are highly dependent on the environmental conditions for their development, and the larvae can remain at a level of dormancy for many years under unfavorable conditions. They tend to run away and protect themselves from insecticides and other treatments that aim at such insects by crawling into tiny spaces. They also prefer lower levels of humidity that are approximately lower than 2% and they have a low adaptation to high humidity. This means that there are increased chances for them to spread in the dry and hot conditions that are found in the Arid Southwest. It is therefore, highly important to take note of this pest and make sure that the necessary quarantine measures are followed in this situation. Human activities are almost always the ones that introduce and spread this pest because the adult hardly ever flies.

9.3.5 Flour Beetles

Above is the common name for *Truboliom* spp. This species is from the order Coleoptera and family Tenebrionidae. The most occurring species are the red flour beetle (*T. castaneum*) and the confused flour beetle (*T. confusum*). These insects can be found in all parts of the world.

These beetles are brown, elongated and flattened, slender, and small with a length of approximately an eighth of an inch. The best way these pests can be identified is through their reddish brown, elongated, flatten bodies. They have tiny punctures on their thorax and head, and lengthwise ridges and punctures between these ridges on the elytra. They generally have rectangular shaped thorax and they lack any teeth on the margins. The antennae of the confused flour beetle have a gradual enlargement towards the tip, ending in a 4-segmented club. Towards the head, the thorax starts getting slightly broader in comparison to towards the abdomen, however they have almost straight edges.The antennae of the red flour beetle have an abrupt enlargement of the tip forming a three segmented knob. It has rounded edges on is thorax, which is slightly wider in the middle (Hagstrum & Subramanyam, 2016).

This pest is important due to its preference for a wide range of stored grains, as well as their products such as flour in cereal. These pests hardly ever destroy full grains, but they are found more in flour and broken beats, or grains that had initially been infested by other insects. The hardly ever cause damage in terms of feeding, however the damage is mainly due to the contamination of the stored product, mainly because they have a number of specific life stages, fecal pellets, cast skins, and dead bodies. During their activities, they also tend to produce pungent substances and these can lead to a serious reduction in the grain quality, therefore reducing consumption by human and livestock as well as marketability. When they infest heavily, they also lead to the buildup of heat in the containers that are storing the food, which leads to secondary mold growth (Hagstrum & Subramanyam, 2016).

Figure 9.7: These beetles can be breeding all through the year in areas ghat are warm. Their life cycle is appointment 40 to 90 days, and the adult has an ability to live for 3 years. It is possible to find all the stages of its life cycle at the same time in infested grain.

Source: https://www.shutterstock.com/search/red+flour+beetle

When it comes to the lifecycle, they loosely lay their eggs on the stored products, and the adults together with the hatching grubs feed together on the products until they are exhausted. They also pupate on the food products and they do not enclose the pupae in cocoons. They are highly dependent on temperature for the duration of the lifecycle, which takes approximately six weeks in temperatures higher than 30°C, and this duration could be much longer in months that are cooler. In a year, they can create up to 10 generations. They have highly active adults that tend to live long, majority of them living up to one year or even more.

9.3.6 Saw-toothed Grained Beetle

Oryzaephilus surinamensis under the order Coleoptera is commonly known as the saw-toothed grained beetle with its closest relatives being the grain beetle and similar in such a sense that they can often be confused for one another is distributed all across the world. It is a small slender beetle that is flattering that the long-awaited with black and brown markings and grows to about 1/8th of an inch in length: its most defining features is a 6 saw-tooth like projection that run along both sides of the pronotum on the metathorax which connects the thorax on the abdomen. The elytrae on this species are elongated with rounded ends and it also has miniature puncture like depressions that cover the abdomen. This species is built with prominent eyes a long antenna and the eggs that lays hatch into lava that are yellowish or white with a brown head. They are built like elongated worms with sparse hairs along their body. Like all insects repairs of short legs can be seen on the thorax. The Merchant grain beetle as mentioned above is almost identical to the saw-tooth grain beetle except for the noticeably large eyes the located a little farther away from the neck when comparing to the saw-tooth beetle. It is an important pest when mentioning storage habitat insect and can be found in a wide range of stored products that have both plant and animal origin.

As already mentioned they can damage a large number of stored products in the pantry and also surrounding areas feeding on the area of stored seeds that in case them and if they find a way to some greeneries they can hack them to such a way that they reduce the germination potential that the seeds have significantly. Grubs and adult beetles are like can- balls through stored products reducing them to find products after totally consuming them. Adult saw-tooth grain beetles have such strong Jaws that can easily chew through plastic packaging and make quick work of paper packaging to reach the food sources. If infestation numbers are unmitigated they can cause storage products to heat up can begin to develop mold which will reduce the quality current flavor of the food items in question.

Figure 9.8: In the house, sawtoothed grain beetle problems start when homeowners insert infested food into their homes. The insects spread quickly to other products that are boxed or bagged. As adults, these beetles may end up leaving the pantry and appearing in other parts of the house.

Source: https://keys.lucidcentral.org/keys/v3/eafrinet/maize_pests/key/maize_pests/Media/Html/Oryzaephilus_surinamensis_%28Linnaeus_1758%29_-_Saw-toothed_Grain_Beetle.htm

The life story consists of eggs being laid loosely in stored products and the lava or grub that are hatched and adults together would all feed on the stored material until it is completely exhausted. Because the lava needs air you will find them within the top two or three inches of the stored material and even the pupation stage of their life will take place within the food product. The occupation stage begins with a mature level which will create a small cocoon with particles of the substrate and their saliva in which they encapsulate themselves in. The duration of their lives can vary and it is majorly dependent on the temperature that exists with the species that are born and live in warmer temperatures above 30 degrees Celsius having between three to four weeks of life and does that are born and live in colder months will live longer. Within a year they can be anywhere between 4 to 6 generations of the species.

9.3.7 Spider Beetles

Ptinus spp, *Mezium* spp and *Gibbium* spp are all names of the Coleoptera Order of insects commonly known as spider beetles. They go by different names depending on where they are found on the earth because the distribution is global. They can be identified by their small globular brown or black physic with some having colored markings or patterns. They will grow between 1/8 to 1/4 inch in length and they can be best featured through their spider- like appearance and their aforementioned globular pear-shaped abdomens that is fully covered with elytra. When viewing the insect from above, their head is pointed downwards almost completely hidden from view. Sometimes they may have a shiny dark reddish color on their abdomen and light brown legs and antennae. The antenna is usually long and slender and so are the legs (Hagstrum, 2016).

White-marked spider beetles grow a little bigger to about a quarter inch and have a light reddish brown color with patches of white hairs on their elytra. You will find that the females of these species have two white patches on each elytron that join to form some sort of traverse bands. The legs and antennae also different in color and will be a little pale yellow. This species will only eggs that hatch into lava Trust more plump and c-shaped. They will be like less on their graves with light brown heads. The adult spider beetles will often be confused with bed bugs but upon closer examination it is quite easy to distinguish the difference. This scavenger is a pest that can be found in a wide variety of stone Products but have a favorite affinity towards cereal and grain stored products. Spider beetles do scavenge feeding widely and when they do decide to infestit is location but we will do well wooden structures exist near the food source.

In this scenario, they can cause quite extensive damage because they go have areas where the lava can score the wood so that they form their pupals self or cocoons to reach full maturity. Luckily they do not bite or sting human beings or pets and they are not carriers of any disease. The life will consist of eggs being laid in stored product or simply on the outside of packaging or debris that is found around storage structures sometimes even in crevices. The larvae will feed on any available food source because they are very good at borrowing especially through wood and even cardboard structure when they need to be paid. Over winter some species of spider beetles will halt their pupation until the following year in the spring. Other store will Simply Be active all year round but you will find that a little longer than the saw-tooth grain beetle with the Year hosting two or three overlapping generation (Hagstrum, 2016).

Figure 9.9: Spider beetles are quite tiny, they are actually under an inch in length. This means it is possible for them to hide in generally any crack or crevice. For example, they normally hid in the cracks between floorboards. It is very way for them to enter packets of food. They chew their way in.

Source: https://www.britannica.com/animal/spider-beetle

9.3.8 Yellow Mealworm

The yellow mealworm is scientifically known as *Tenebrio molitor* and fall under the order of Coleoptera making them beetles in essence. They can also be found all around the world and that is their distribution and they consist of some of the largest -stored product group of insects. The adults are usually Lord shiny dark brown or black beetles that will grow to half an inch and sometimes even three quarters of an inch long and they will we often be seen in the early summer flying into light. The most definitive features are the adult thorax that is covered with small little punctures and their elytra having longitudinal grooves running along them. The larvae of the yellow mealworm are ordinarily slender and pale -yellow and will start turning dark yellow when they begin to mature. Each segment of their body is covered with a yellowish-brown border (Hagstrum & Subramanyam, 2016).

Like all stored product insects, they are pests but occasionally they will be found in stored grains and other store products. New ones will feed externally on store grains but do not actually burrow or bore into them or cause any webbing so they are not considered serious pets as such. We will however cause damage of a considerable amount if all the grain is stored on a layer for example and we'll end up contaminating stored products with its feces and cast skins when they molt. Severe infestation can be quite extensive in the destruction levels especially when the stored product is left unchecked for long periods of time. You will find that their life cycle will begin with eggs being laid and glued onto the food source in which they will hatch and begin to feed. The larvae feed continuously through the summer and this stage and most times able over the winter as larvae. They undergo pupation during the springtime for around two weeks, emerging as full adults usually in the late spring or early summer and because of this particular cycle, you will find only one generation in a year.

9.4 MOTHS

9.4.1 Angoumois Grain Moth

The above is the common name for *Sitotroga cerealellea*. This species comes from the order Lepidoptera and family Gelechiidae. It is found in all parts of the world. The adult months are straw or buff colored, have a dark tan, are slender, and are smart, with a length of about a third of an inch and a half of an inch in the wingspan.

Figure 9.10: Females normally end up laying up to 160 eggs, either one by one or in clusters, on the outside of grains. The Larvae normally burrow into the grain's inside where they now feed and remain until they emerge as adults.

Source: https://www.grainscanada.gc.ca/en/grain-quality/manage/identify-an-insect/primary-insect-pests/angoumois-grain-moth.html

The main features that can be used to identify them include the fringe of hair found on the back edges of both pairs of wings. They also have finger-like projections on the ends of both pairs of wings, with the hind wings being more distinct in this feature. They have creamy -white larvae that have black heads. They also produce reddish- brown pupae which are found inside cocoons that they make out of silk. Grains that have been damaged have circular holes that are the same as those seen with stored grains that have beetle pests such as the maize weevil and the rice weevil. It is important to locate the adults so as to discover where the source of damage is.

This moth is an occasional pest of stored grains as well as others toward products. It also tends to attack corn which is still standing in the field. All the damage that occurs due to this moth is fully due to the larvae. The end up boring into the grain's kernel and spend the whole pupation and larval stage inside it. Kernels that are damaged are hollow and majority of them have a circular exit hole from which the adult used to get out of this grain. Grains that have been infested do not germinate. They have a smell that is unpleasant making them unfit for human or livestock consumption. The moisture content and temperature of stored products can be increased by infestation, leading to the attraction of mold and other microorganisms such as bacteria. The adult moths do not damage nor feed on these products;

however, they only go to these products to lay the eggs. Stored ear corn has a higher level of susceptibility to damage from these moths in comparison to shelled corn. Before the grains are harvested, infestation can always start in the field, or it can be gotten from stored grains that were already existing (Hagstrum, Phillips, & Cuperus, 2012).

The adult months normally lay and glue their eggs to the surface of the grains, majority of the time while the grains are still in the field. After the larvae hatch, they immediately bore into the kernel of the grain and seal the hole from which they entered. They spend the rest of their time inside the grain eating and hollowing out the content as they go through about three different molting stages. When the larvae are mature, they chew a round hole on the kernel surface and thereafter use silk to seal it and then pupate within the grain in a silken cocoon. When the moth is now an adult, it emerges and flies out of the cocoon through the hole it had created to exit, and the cycle repeats. In one year, these insects could have four to five overlapping generations; in favorable conditions, many more could occur.

9.4.2 Indian Meal Moth

Above is the common name for *Plodia interpunctella*. This species is from the order Lepidoptera and the family Puralodae. It is distributed all over the world. The adult moths are shiny and small, with a length of approximately a quarter of an inch and ¾ th of an inch in wingspan. The adults are more commonly seen and they also have an attraction to light.

The main identifying feature is very distinct in the newly emerged moths, where their thorax and heads are reddish brown in color. They have a tan or light gray color on the top half of their forewings and a coppery border with dark stripes or reddish brown in the bottom half. They have fringes of dark gray hairs on the edges of their forewings. It is only during flight that you can see their hind wings, which are tan or uniformly light-gray. The majority of the moths as they grow and live on tend to lose their characteristic colors because they come into contact with many surfaces, therefore appearing gray or uniformly tan in color. They have a pointed head, with long slender antennae and prominent eyes (Hagstrum & Subramanyam, 2016).

Figure 9.11: This species was given its name after being noted for its behavior to feed on cornmeal or Indian-meal and it does not natively occur in India as would be suggested by the name Indian mealmoth. It is should also be treated differently from the Mediterranean flour moth (*Ephestia kuehniella*), which is similar and also a common pest of grained that have been stored.

Source: https://en.wikipedia.org/wiki/Indianmeal_moth

They normally slightly raise their forelegs when they rest, making the body have a sloping appearance. The larvae have variations, being either light green, pale pink, pale white and sometimes even brown in color. They have a length of approximately ½ an inch. On their abdomen, they possess five pairs of prolegs that are short and stubby, allowing them to easily move around from within their substrate. They have brown colored pupae that have a length of approximately ¼ to ¾ of an inch, and they may be found inside or without a silken cocoon. Despite the fact that the adult moths do not damage or feed on the stored products, their presence can be the sign of an infestation. The larvae can travel away from their source of food, even having the ability to climb walls, to various places for pupation, and when trying to find the source of infestation, this can be misleading (Hagstrum, 2016).

This is a pest of many different stored products. It is the commonly seen moth pest of stored grains. The larvae entirely cause all the damage, by feeding on the source of food, and also the fact that they spin many silken webs. They build these webs to prevent them from desiccating, however, they attract cast larval skins, debris from the damaged food source, and fecal pellets, all which can lead to a higher level of contamination than feeding on the grains themselves. These silk webs can be cumbersome during the time the grains are being processed, because they tend to clog the mill machinery. The temperature and moisture content can also be increased by high levels of infestation, all which can attract mold and other microorganisms. One can find most of the larvae in the top two to three inches of the stored product. When the packages are closed, moths cannot enter. They require an opening such as a seam or a hole to access the source of food so at to lay their eggs. Adult moths normally feed on nectar therefore being very unlikely for them to destroy the stored products.

When it comes to their life cycle, two to three days after the adults emerge, they mate, thereafter laying their eggs directly on the source of food. The larvae then proceed to feed inside the source of food, however, they also have to ability to move to another source nearby. They could also pupate in the source of food or in a different location. They highly depend on temperatures for the duration of their life cycle which is done in approximately four to five weeks under optimal conditions. In a year, they can have several overlapping generations.

9.4.3 Mediterranean Flour Moth

Above is the common name for *Ephestia kuehnielle*. This species is from the order Lepidoptera and family Pyralidae. It is distributed all over the world. The adults are shiny pale gray, slender, and small, with a length of approximately ½ an inch, and ¾ to 1 inch in wingspan. They have small larvae that are pinkish or whitish in color, having a black head and tiny black spots covering their body. They pupate in cocoons made of silk, many a times a distance from the source of food. They have reddish brown pupae.

The best way to identify them is through their forewings which are a pale pink-gray having a number of dark spots, two broken zigzag lines towards the ends and a broad dark band towards the middle. They have uniformly white or pale gray hind wings, having a thin dark boarder around them. They normally slightly raise their legs when resting, giving a sloping appearance to their bodies.

Figure 9.12: The best way to control this pest involves having strategy for sanitation of sealing grain containers and facilities to avoid them from infesting the grains, however, it is also possible to use some pesticides.

Source: https://en.m.wikipedia.org/wiki/Mediterranean_flour_moth

They are occasionally found around stored grains and other products. The larvae are entirely responsible for their damage, mainly because they feed on the source of food, and spin silken webs. The webs add to the contamination together with cast larval skins, debris from the damaged food source, and fecal pellets. There can also be an increase of temperature and moisture content of the stored products with heavy infestation, which leads to the attraction of mold and bacteria. They can also be highly cumbersome when the grains are being processed. These moths are also not able to enter packages that are closed, and require for either a seam or hole as an opening to access the food source but only to lay their eggs. Just like the other moths, they feed on nectar and never damage the food.

The female moth normally lays her eggs directly on the source of food. Immediately the larvae hatch, they start spinning their webs and feeding voraciously on the source of food until they have reached the stage of pupation. When the larvae are mature, they leave that specific location next to the source of food to pupate. As with all the pests of stored products, the life cycle is also dependent on temperature, and is done in approximately seven weeks under optimal conditions. In a year, they can have four to five overlapping generations (Hagstrum, 2016).

9.5 CLOSET PESTS/PESTS OF FABRIC

9.5.1 Carpet Beetle

The Antherenus or Attagenus in the order of Coleoptera is commonly referred to as the carpet beetle that are fabric pests of different varieties inhabiting the furniture carpet and are most commonly found in the South West contiguous United States although their distribution is global. The physiological descriptive features include a small dark rounded or oval shape and build that grow up to 1/8th of an inch in length some which would grow a variety of patterns mainly resembling lady beetles. The larvae that hatch from the eggs of the carpet beetle are usually short and plump with a worm-like appearance and they are more often covered with long hairs. The most descriptive feature that can easily identify the carpet beetle is a shiny black or dark brown hue with brown legs and the lover of this insect would have a light or reddish- brown tone as well growing to a quarter of an inch with the biggest extending to maybe half an inch long and they appear much bigger than what the adults are. The rest of the body is also covered with short and sparsely distributed hair.

Furniture carpet beetles that are scientifically known as *Antherenus* exhibit black white and brown patterns with tanned scales. As the beetle ages, these scales will fall off with the older adults having an almost uniform dark brown or totally black underside with white markings. As described above the larva of this insect is pale yellow but as they grow older as well they change turning dark brown with distinct markings forming on them of light and dark stripes. Towards the head of the larva, is a little thicker with a tapering happening towards the rear end with tufts of hairs. The rest of the body though is covered with dense hair as well but it is more noticeable at their tail.

A variety of carpet beetles will look very similar to furniture carpet beetles with the distinctive black white and brown patterns with some yellow undertones although these markings will fade as they age into mature adults and becoming more uniform in the brown or black gloss. Their larvae though are almost indistinguishable from the furniture carpet beetle larva in that they are plump and dark yellowish to brown grubs however their shape is different in that there are broader towards the rear end even though they are still a little bit larger than their adult counterparts. If or when disturbed they tend to react by holding the hair up right and freezing making them appear to be larger than they are and more like a small clump of fibers. The larvae are also annoying in that they leave numerous cast skins when they molt and will also shut off the bristle or has that they have.

The damage that they produce can be quite large and appear to be continuous and this is important in that it can help you distinguish when the damage was done by carpet beetles or clothes moths. Clothes moths and the lovable leave behind a thinner and smaller cast skin together with weddings and closes and this will come in the form of small holes that is well distributed across the fabric in question though not as extensive as those produced by the carpet beetles.

Figure 9.13: The adult carpet beetles normally fly inside through windows and doors that are open so as to lay eggs on rugs, clothing or furniture, and often are introduced from bringing home infested items. If items are not kept clean and well maintained by homeowners, a new brood of these beetles may hatch thus starting the cycle again.

Source: https://www.orkin.com/other/beetles/carpet-beetles

They are considered moderate pests in that way produce damage on all kinds of fabrics and similar items although they will be really found in stored food product on which they can cause allergic reactions to those who may be sensitive. Damage or injury is usually caused by the larval when it comes to feeding on stored products and they will attack a variety of fabric and other similar items but they usually have a higher affinity towards those made from animal products like silk wool leather all kinds of furs, feathers and even taxidermy mounts. They will also attack all sorts of fabrics made from plants like cotton and jute as well as different kinds of upholstery carpet tapestries and such. Some fabrics that are blended with both natural and artificial products will also be subject to attack by the furniture carpet beetle but purely synthetic or artificial fabrics like polyester and nylon are not commonly found damaged.

When these insects decided to go to work it will most commonly occur during the dark in hidden and undisturbed areas such as cracks and crevices in the furniture or inside the rules of carpets and rugs or various foods in clothing materials. Various food substances are also not discriminated by the larva of furniture carpet beetle. It is not uncommon to hear that people will have allergic reactions to the larvae and the hairs or bristles that they produce often induce an intense irritation or itching around areas where they come in contact with said bristles with presentations going so far as to affect not only the skin but the respiratory tracts of the subject period in most cases long-term exposure is what would result in individuals having these sort of reactions because of heightened sensitivity.

The adult furniture carpet beetles do not feed on fabrics or stored food products like the larvae do but instead choose to fly outside where they prefer feed on plants especially those with a good amount of pollen and nectar from which they will be transferred back indoors with the flowers that people will pick. The only time that furniture carpet beetles or carpet beetles ,in general, will come indoors is when it's time to lay eggs and if they cannot find access to a closed environment, they will invade outdoor habitats such as nests of birds and other small animals that may shed fur. Most infestations in doors begin with the adults being brought inside with infested material like the aforementioned animal and plant fabrics or cut flowers. Once they find a suitable location inside, they will fly around the home or the building looking for a place to lay their eggs and they are very good at finding air ducts, closet and sneaking on the furniture to do so.

The life history of carpet beetles includes eggs being laid directly on food sources which is the fabric material often preferring woolen or carpet and first and when the larva hatch they continuously feed especially in dark and undisturbed areas with an insatiable appetite. The larval periods of this insect can vary with some extending even up to three years but in favorable conditions they will be done growing in about 3 months entering the pupation stage. At the entrance stage, they will often be moved away from the food source so that they will not be cannibalized by their peers. Many of them will over winter or become totally dormant just before they pupate. If there is no suitable shelter away from the food source you will find that the pupation stage will occur near the food source under a concealed area or even under the last larval skin. They will remain here for about until the late summer or early fall when they emerge as adults leaving for 2 weeks simply to mate and reproduce finding suitable locations to lay eggs. In a year, there may exist 2 generations that overlap but they do not tend to live very long beyond that.

9.5.2 Clothes Moth

Above is the common name for the *Tineola* spp and *Tinea* spp. These two come from the order Lepidoptera and family Tineidae. The most common are the case-making clothes moth (*Tinea pellionella*), and the webbing clothes moth (*Tineola bisselliella*). These have a worldwide distribution. They are shiny yellowish- gray in color, slender, and small, with a length of approximately ¼th of an inch, and ½ an inch in wingspan. These are much smaller than the ones discussed above as stored food pests.

The webbing clothes moth can be identified with its yellow color and sheen that is distinctly golden. Their wings also have hairs that are golden. It has reddish- brown hairs on its head. For the case-making clothes moths, they have more of a silvery or graying color, with spots on their wings that are faint and dark. They also have yellow hairs on their head. For both species, the adults do not like to fly much, and they also do not like light. Many a times, they are found in dark places, areas of the closet that are undisturbed. Normally, they escape by running when they are disturbed instead of flying, and if need be, they flutter.

Figure 9.14: These moths rarely ever attack clothes and blankets that are used regularly. They also do not attack carpets and rugs that are routinely vacuumed or have constant traffic. The most affected are sections under the furniture and the under sides of rugs.

Source: https://entomology.ca.uky.edu/ef609

In both species, the larvae look similar, being cream colored or small white caterpillars with dark brown heads and the length of about ½ an inch. We mostly find webbing clothes moth larvae under webs that they have created with silk, feces and fabric debris. We then find case-making clothes moths in a case that is silken and spindle shaped. They carry this case around as they are moving. The larvae normally leave behind delicate small cases or webs, and cast skins. They also leave around tiny holes that are widely dispersed on the fabric. This damage makes it easier to distinguish them from carpet beetles, which normally create large contiguous areas of fabric that is damaged and cast skins that are shell-like and covered with long hairs or bristles.

These pests are occasionally found in fabric and items similar to fabric, but they are hardly ever found in stored food. The larvae cause all the damage as they feed on the source of food. They attack all fabric types, including tapestries and carpets, upholstery, taxidermy mounts, bristles and brushes, felt, hair, feathers, furs, wool, silk, cotton, and similar materials of plant and animal origin.

Their most preferred food is wool. They also attack blended fabrics that have wool, silk or cotton along with synthetic fabrics, however, they do not attack fabrics that are purely synthetic such as nylon and polyester. They mostly choose hidden spots to damage, such as cuffs and under collars, rolls of rugs and carpets, crevices and cracks of furniture, and hems and folds. Fabrics that have stains of food, urine, or perspiration always have higher chances of getting attacked. The adults never feed, and their main purpose is just reproduction. They only use the stored products as surfaces for laying their eggs, and never lead to any destruction. It has also been noted that clothes moths may also live in bird nests, feeding on cob webs and other fibers.

In their life cycle, soon after they emerge as adults, they mate. And the females end up laying clusters of eggs, which they use a glue like substance to stick to the source of food. Soon after they lay their eggs, they die. It is more likely that the males live for about a month longer than the females. When the larvae hatch, they immediately start feeding, and the two different species make their own webs and cases. The cases and webs are normally the same color as the substrate, because they use the substrate's fibers. As they move about, the webbing larvae normally make extensions of their web tunnels, and there is accumulation of fibers, debris and feces within their webs. This is different in the case-making larvae that use their case all through their larval stage, and as they grow, the case increases in size as well. They normally have their case with them wherever they go as they are moving on the substrate. When they need to feed, they extend their heads from either side of the case. The case-making larvae normally pupate in crevices or cracks away from their feeding sites, while the webbing larvae pupate on the source of food in silken cocoons. They are dependent on temperature for their life cycle to occur, which is normally done in four to six months. In one year, they can get up to two generations.

In general, if one notices any of these pests in their homes or in the storage facilities, immediate action needs to be taken to avoid the hatching of a new brood.

10

Desert Insects

CONTENTS

10.1 INTRODUCTION

In Deserts, just like other lands around the world, insects account for higher numbers than other types of fauna altogether. Desert insects feed in different ways, some slice root, leaves, and stalks into shreds, others pounce other animals, consume waste and so on. They also possess the ability to breed prodigiously. For example, a queen termite, which is among the highly productive insects, might produce millions of young ones during its lifetime. Luckily, there are certain hostile situations, parasites and predatory opponents that inhibit such factors from occurring and therefore the bugs are kept in control.

By any standard, insects have a big part to play, across different food-chains, in terms of influencing desert patterns. They have developed into herbivores, omnivores, and carnivores, some insects are greatly specified and others rather general in their food selections. Being tiny animals, they also can explore environments more completely compared to larger animals. One plant might be food for vertebrates, however, for arthropods (which comprise of some bugs) it could be a whole world. A single desert insect type may pass through larvae growth within a floral bud, whereas a different species may spend its whole lifetime consuming the woody stems.

10.2 COMMON TYPES OF INSECTS FOUND IN DESERTS

10.2.1 Sand Cockroaches (Arenivaga)

These are native to arid areas and are also called desert cockroaches. Their main habitat is underneath the surface of sand. Sand cockroaches are huge insects that exhibit substantial sexual dimorphism, meaning their males and females are unique in appearance and might not seem to be members of one particular species. The male sand cockroaches are typically winged and look like American cockroaches or cicadas, but only lighter in shade. Males are often seen in flight. Female sand cockroaches meanwhile are wingless and darker in shade than males and have a unique oval exoskeleton or external shell.

Figure 10.1: Sand cockroaches (Arenivaga).

Source: https://en.wikipedia.org/wiki/Arenivaga

Female sand cockroaches appear less like the typical cockroaches we're used to watching and more like primeval (and extinct) organisms called trilobites. Male sand cockroaches are usually attracted to light, and most probable to be seen in daylight hours, whereas female sand cockroaches are largely nocturnal coming out at night (Tarbox, 2007).

Sand cockroaches regularly feed on desert plant roots and leave their tunnels infrequently. They also breed inside their burrows and aren't known to invade homes. Persons who suffer from certain allergies and/or asthma must avoid any contact with sand cockroaches, irrespective of where they find them.

Arenivaga, a species that formerly held nine species now has been found to have close to 39 different types. The Corydiidae group of roaches is found globally and its constituents are regularly found in harsh, dry habitats not typically associated with cockroaches. They are furthermore found in subterranean zones, and their habits make their presence rather easily overlooked. These insects have amazing adaptations that permit them to thrive in some of the toughest places on earth. The application of modern collection procedures has revealed many species of *Arenivaga* around Mexico, including many other species of Corydiidae found in deserts and dry regions of the rest of the globe (Tarbox, 2007).

Arenivaga are also dimorphic, meaning the females look absolutely nothing like their male counterparts. This makes the identification of female specimens of the species against male specimens of the given species rather difficult. As part of the order Blattodea (cockroaches), they comprise some of the globe's greatest decomposers. They are the earth's recyclers and evidently they have a part to play even in land with minimal plant matter needing decomposition. Cockroaches are no longer thought of as creatures limited to the humidity of the tropics, desert cockroaches after all survive in arid regions (Tarbox, 2007).

10.2.2 Desert Bees

Many desert bees roam alone, even though they regularly assume dwelling in bee-suitable environs. Normally, a female bee digs a private tunnel some few centimeters to 6 feet down in the dry sand. Lining up the tunnel with tissue, which she protects using a wax-like emission from their stomach tissue. The bee then supplies every cell with fresh nectar and flower pollen, laying her eggs in every tissue.

Figure 10.2: Desert Bees.

Source: https://www.youtube.com/watch?v=r-6B2f5XRe0

Contented that her labor is thorough, the female bee leaves the tunnel, sending her brood away to fend for their existence for the remainder of their existence. Some desert bees, instead of excavating tunnels, go into deserted channels left behind in dead timber by lumber-digging beetles, besides they make their tissues from foliage pieces, tiny gravels and mud.

Some few types of desert bees, same as their close family, the ants, have social lifestyles. There are furthermore natural opportunistic bees that designate the provisioned layers of other hardworking bee forms, plus they produce eggs that mature very early feeding on the available provision of pollen or nectar as well as the larval phase of reluctant hosts.

10.3 HOW DESERT INSECTS HAVE ADAPTED TO SURVIVE THE HARSH ENVIRONMENT

Figure 10.3: A desert insect conserving water on its body.

Source: https://www.dw.com/en/survive-and-thrive-in-the-namib-desert/g-39542650

The two key adaptations that desert insects must make are dealing with lack of water, plus how to deal with excesses in temperature. Many desert insects avoid the day heat of the desert by basically staying out of it as often as conceivable. Since water is rather scarce, most arid region insects derive their moisture from nourishment they consume such as juicy plants, seeds, and tissue matter from their prey for the case of predatory insects (Weber, 2003).

Some inhibit water from leaving their bodies by diverse means, like living in tunnels that don't become extremely hot or cold plus have moister (humid) air inside. They live in their dens during the hot days while coming out at night to feed. Other insects have bodies designed to conserve water, while some have a thick exterior covering which lessens moisture loss. The kidneys of desert insects concentrate urine, so that they expel less water.

One cool insect adaptation is the Namib Desert Beetle's inventive water collection. The beetles exist in an area with just a centimeter of rain per year, but the bumps found on its back cause liquid in the air to gather on it. To get moisture from the air, the beetle uses its elongated legs to place its body at a 45-degree position, head downward, facing into a calm early-morning wind. Small water droplets develop on its bumps, which are covered in hydrophilic compounds that attract water.

When enough water gathers to make a droplet, it cascades down the water-resistant passages between the bumps, and into beetle's mouth. Insects have adapted to survive the harshest desert regions, besides other animals like scorpions and some reptiles. In deserts with greater humidity, the level of biodiversity surges as vegetation like shrubs, cactus and hardy trees are the foundation of a more elongated food web. As deserts are present in most continents, diverse species have progressed to survive the harshest desert environments (Weber, 2003).

Some insects have humps to store fat which are broken down into water and energy sources when nourishment isn't available; they rarely sweat, even in hot temperatures therefore when they do consume fluids they can preserve them for extensive periods of time; Besides, they have huge, sturdy lips that permit them to consume dried and sometimes prickly desert foliage; They may also have spiny, level, feet to spread out their weight and offer protection from the hot sand during movement.

10.4 WATER SOURCES FOR DESERT INSECTS

10.4.1 Dung

While it may seem disgusting, dung is actually a reliable and convenient source of water for many desert insects like beetles. Although it might look gross, consumption of dung is a great adaptation for tiny desert insects. In hot and dry areas, moisture is quite low and difficult to come across. Dung has water from the stomach of creatures that expel them.

Rather than looking for scarce water holes just like other larger animals do, these insects wait for their bigger animal counterparts to perform the task of finding fresh water for their consumption. By eating this dung, the insects derive all the benefits of water discovered by others without having to carry on any of the hard task (Wehner, 2019).

It doesn't mean that such insects live in leisure. Many types of beetles, for instance, spend long periods shaping the dung into flawless spheres, which they then roll around the desert into their dens. Based on the scope of the dung, it may offer sufficient nourishment and humidity to keep the insect active for more than even 1 week (Wehner, 2019).

10.4.2 Rainfall

Many insects are active during dawn up to dusk when temperatures become much cooler. During the peak of noon, they dig into the sand to escape the high temperature. Their glossy exoskeletons imitate sunlight, which inhibits them from getting too hot. Insect life in deserts is a 'continuous battle to covert moisture into subsistence.' Water has conventionally been a protective habitat for insects, but rainfall is scarce and desert storms are quite violent. A record 44mm of rain once dropped in just 3hrs in Sahara Desert. Huge Saharan storms meanwhile can deliver up to 1 millimeter per minute (Tarbox, 2007).

Typically, dry tributaries, known as arroyos or wadis, may rapidly become filled after heavy showers, and flash floods make the waterways rather dangerous. More individuals drown in deserts than perish from thirst just for comparison. Though minimum rain drops in the deserts, they still have runoffs from ephemeral, or momentary, streams fed by rainfall or ice coming from neighboring highlands. These creeks fill up the waterway with a slurry of mud and normally transport substantial amounts of residue for one day or several.

Even though most arid areas are in basins having closed, or internal drainage, some few deserts are traversed by 'exotic' rivers that get their waters from out of the desert. These rivers permeate soils and evaporate huge volumes of water on their course through the arid land, but their volumes are so that they preserve their continuity (Tarbox, 2007).

10.4.3 Temporary Desert Lakes

These lakes are normally shallow, brief, and saline. Since the lakes are low and with an equally low base gradient, wind stress can cause the lake waters to transfer over many kilometers where more insects can have access to it. When small lakes become dry, they often leave behind a saline crust or hardpan. This level zone of clay, silt, and sand covered with salt that develops is called a playa. In North America alone, there are well over such

100 playas. Many are relics of huge lakes that existed during the previous Ice Age roughly 12,000 years back. Insects not only come to these playas to hydrate after a downpour, but also feed on their rich salts and minerals.

10.4.4 Oases

Oases are frequently the only reliable places in deserts which can support permanent insect habitation. The term 'oasis' is believed to come from the Egyptian term wah, meaning 'fertile place in a desert'. Oases are typically areas in a desert where subsurface water comes to the surface due to natural springs and human-made wells. They are typically found in low or central areas of a massive depression or inside fissures of a plateau, typically ringed by sheer sandstone cliffs.

Sometimes, these oases are quite large thus, capable of supporting different types of insects at once. One oasis can be several miles long and encompass many villages. A classic oasis is a basin having fertile soils at their core deposited from wadis. Groundwater is relatively near to the soil surface, where it sprouts and nourishes insects living in the surrounding area.

10.4.5 Springs, Wells and Temporary Streams

Springs and wells usually are served by sporadic rains and subterranean penetration of runoff water from slopes and mountain ranges. Capillary activity springs up subsurface water from water tables. Water present in desert wells is frequently brackish, making it the ideal spot for mosquitoes and other insects that prefer stagnant water.

In some regions, especially in and surrounding oases, water is near enough to the ground that it may be carried up by fairly superficial wells. In many of the arid areas, there's groundwater however it may be far below the ground and only extracted at immense overhead cost, using drills for digging deep wells and pumps for bringing it up.

Some deserts that receive minimum or no rain at all receive water from the rivers. Among the highly populous desert ranges are regions where water is drawn in by rivers originating from outside the desert. That's the situation with the River Nile in Egypt as well as Sudan, Tigris and Euphrates that passes through Iraq and Syria, as well as the Indus River of Pakistan. Wadis are periodic ricers which are dry for most months and jam up with water in rainy seasons. Water may typically be extracted in wadis, plus at the top of alluvial fans whenever no water is available at the surface. The presence of plants is frequently an indication the water is near enough to the ground that a well may be effortlessly dug.

10.4.6 Aquifers

They are underground films of porous rocks containing water. Shallow aquifers may easily be replenished through rainwater. Deep aquifers can't be rapidly replenished. By some estimates, there's roughly 100 times extra subsurface water than there's ground water in the lakes or rivers. Drawing too much water from shallow aquifers can cause environmental destruction, especially when this is done artificially. However, sometimes the water is pushed up the surface from underground pressure where insects can have access to it. Deep artesian wells rupture up where rock structures conserve water under immense pressure. For some oases, an artesian spring drifts to the surface from a crack found in the ground.

10.5 HOW DESERT INSECTS EVADE PREDATORS

10.5.1 Camouflage

Figure 10.4: A desert grasshopper camouflaged to resemble the surrounding rock.

Source: https://www.pinterest.com/pin/319403798544618448/

Insects often hide against a similar-shade background. These might be green or tan so they conform in with their surroundings, or they might change pigment slowly to conform to variations in their immediate background. Insects likewise use other trickeries like changing the shape of their bodies, and evading rapid movements that may expose their location. Undercover desert insects meanwhile use disguise. This involves resembling another object, permitting the camouflaged insect to hide right in plain sight. Also, stick insects like the walking stick resemble twigs or branches, whereas leaf insects like the katydid look like the leaves they consume (Weber, 2003).

For desert butterflies and moths, they have eyespots around their wings imitating the expression of an owl or some other big predator. These eyespots may be shown to frighten a bird or some other potential predator, providing the insect an opportunity to escape. In another ruse, the insects resemble dangerous, foul-tasting or toxic insects to dupe their hunters. In particular, the elephant hawkmoth caterpillar, as well as the viceroy butterfly are among the tricksters who apply this trick for duping unsuspecting hunters.

The Walking Stick insects resemble twigs and branches, permitting them to become imperceptible simply by staining motionless. They might walk sluggishly or sway for purposes of mimicking a branch wafting in the breeze. Stick insects are typically brown or green in color and vary from half an inch to 2 feet long. They freeze whenever they feel endangered. Stick insects aren't constantly as innocent as they look, though. The desert stick insect sprays a moderate acid for warding off hunters. If the acid enters your eyes, it can sting and even momentarily blind you (Weber, 2003).

The wings of most katydids resemble a flowering crop leaf, permitting the insects to hide in open sight in the vegetation they consume. They might also sway back and forth imitating a leaf wafting in the winds. Their camouflage allows them to elude predators, such as fowls, amphibians, snakes, farmers and even gardeners. Due to their brilliant camouflage, it's not easy to see cricket-like katydids, however you can hear them. Katydids brush their wings together, producing the characteristic 'katy-did' sound which provides them their name.

Once katydids are disrupted, they fall onto the earth and exhibit their distinctive color-banded abdomen. This sudden defensive action may scare a bird, which gives the insect a chance to escape. The moth is also an expert in insect camouflage. The color and pattern of their wings match the bark or leaves, and the places where they often visit. The gray-green *Eumorpha pandorus* moth in particular, seems to be in army camouflage attire. Studies have shown that some desert moths are actively looking for the ideal places to live in trees and to find the perfect body position for making themselves undetectable.

Grasshoppers are herbivores that feed on vegetation and sometimes destroy crops and gardens found in desert areas. Grasshoppers on the ground are typically grayish or tan, so they can mix into dry leaves or rock. Grasshoppers living in the grass are shades of green, matching the foliage of their plant loft. Both grasshoppers and locusts have a broken line, and if one leg is wrapped around the clutch of the predator, it can easily break. This lost leg will not develop back, but maintaining a slight disability is a minor cost of survival (Weber, 2003).

The grasshopper freezes when it sees a predator and stays still. Some are boldly painted in red, yellow and black to caution predators that if they are threatened, they shall discharge harmful toxins. These poisonous hoppers raise their wings and wave their legs and wheeze to fowls, frogs, spiders and any other possible predators, trying to make them flinch. At the same time, "flower mantis" is a mantis that imitates a flower to entice and eat insects close to pollen. Scientists studying Malaysian mantis discovered that the shade of the mantis is the same as the color of most Malaysian flowers. They furthermore established that orchid mantis enticed more insects than the real flowers managed.

The insect larvae or Caterpillar are also chiefs of camouflage. A few young caterpillars in the desert imitate bird stools on the leaves, while older bright green caterpillars stay hidden in the leaves. The caterpillar of the hawk moth is grayish-green or brown and has a pair of black eyespots around its head. When vulnerable, the caterpillars will swell up and display fake eyes, making them resemble snakes. It furthermore elevates its head and transfers from one side to another.

The Viceroy Butterfly replicates the look of its prototype, the Monarch Butterfly, for deterring predators. The butterfly imitates the monarch's amber wings since the amber warns the hunter that this butterfly is poisonous. Creatures that eat monarch butterflies will get sick, vomit and remember to elude this or similar butterflies. Both the Viceroy and the Monarch Butterfly transmit toxins and are resentful of carnivores. Thus, any predator that makes mistakes to eat these two insects learns not to do that mistake again (Weber, 2003).

At first, it was believed that the Viceroy Butterfly only imitated the Monarch Butterfly's amber wings to help itself survive but scientists now know better. It has been found that the Viceroy's Butterfly is poisonous. Due to the complexity of nature, the viceroy and monarchs are imitating one another to aid their unique species to survive.

Desert Insects Using Prey Carcasses for Camouflage: Members of the Reduviidae insect family are found in the deserts of East Africa and Malaysia. These insects, also called assassins, pile dead ant carcasses on their backs to obscure predators. Imagine that you find strange things crawling on the leaves, while wandering around their habitat. It seems like a dozen ants, glued to a ball. But look carefully, you will find that these ants are dead. There's a sly insect below, dragging these ant bodies same as a small backpack. Like the bugs of other assassins, it is *Acanthaspis petax*. The insect hunts prey by piercing the prey with a mandible, injects a paralytic agent to induce saliva plus an enzyme that liquefies tissue, and then sucks out the internal organs.

But unlike other insects, it then shapes the hollow ant exoskeleton into a protective coat. This insect can move up to 20 dead ants per time and combine them with viscous excreta to form a cluster that might be bigger than its body. For many years, researchers have been arguing about the abnormal behavior of *Acanthopanax bark*. It hunts many different kinds of prey, however, it seems to only pile ant bodies on the back. Some people believe that the ant's body may provide an olfactory disguise when hunting, whereas others believe that the heaps on the body can be applied as optical disturbances for large creatures that kill these assassin bugs (Auch, 2003).

In 2007, a group of scientists from New Zealand conducted an experiment to determine whether the insect's corpse carrying approach helped guard it from predators. In this research, they left the assassin bugs in a glass cage with some jumping spiders, which make up their natural enemies. Some insects had ant corpses on their backs which researchers marked, while others remained naked. Because jumping spiders have exceptional eyesight but the poor smell, they will use their keen visual sense to make accurate jumps and pounce on their prey when hunting. The experiment was to show whether the ant's body had played as an effective visual camouflage or not.

The result is that spiders preyed on bare insects about ten times more regularly than covered insects. The scientists further repeated the test with lifeless, conserved assassin bugs to manage the effects of locomotion and behavior, where the results remained unchanged. Facts have proved that carrying that dead ant ball is a good strategy that allows the assassin desert bugs to survive the next meal (Auch, 2003).

Scientists believe that massive ant bodies alter the insect's visual appearance so that the spiders cannot identify them as prey. Scientists propose that *Acanthaspis petax* may essentially rely on the spiders' intrinsic hesitancy to attack ants. Since ants possess the tendency of swarming, they may harbor chemical armaments, so spiders typically do not attack them.

10.5.2 Nocturnal Existence

In the insect kingdom, staying awake during the night while sleeping in the day is rather common. Although this may seem strange to most people, many benefits make night action a wise evolutionary adaptation. It's not the insects that decide to change their sleep cycle one day. To be honest, it is nearly impossible to tell the cause of a given natural feature.

The wide majority of desert animals at night are insects. One of the main theories about why they're nocturnal is that the trait evolved into a way to evade being consumed by apex predators such as mammals and reptiles. Although many nocturnal creatures have enhanced senses of sniffing and hearing, it is still simpler to evade detection during the night. Darkness helps hunters and prey travel more invisibly. Prey insects use night cover to seek food more safely, whereas predators use the same night to waylay their victims more effortlessly (Auch, 2003).

For predators that prey on small insects, there is no ideal time than the night outside. If the prey is alert and moving frequently, the predator is usually more effective, so many hunters stalk at night to equal the prey's schedule. There are fewer competitors at night. Besides, some nocturnal species may have developed this approach to diminish conflicts around food sources. In case other animals nearby are going to the puddles and pastures during morning hours, then night movement may be helpful for insects to avoid coming into contact with them.

It is also cooler at night. The nocturnal nature of many desert insects is obvious, for clear reasons, because it can help them evade the noon heat. In a hot ecosphere, moving at night is just a great move. Saving water is likewise an essential aspect of evading high temperatures during the day. Walking around in the shade can avoid overheating and waste of valuable water.

Night insects also have unique adaptability to allow them roam around during the dark. Unlike people, they usually do not focus on vision. In its place, they use sensitive ears and whiskers to perceive adjacent objects and smells and move around. Of course, a number of animals that are active at night also have massive eyes that are particularly suitable for performing well in low light situations. This is particularly true for insects and nightly primates, who must be able to view the fallen branches or leaves (Auch, 2003).

Light contamination is a great risk to nocturnal insects. City lights can cause alterations in migration trends and may obscure insects. For migratory insects like locusts, light pollution can be a major problem since they utilize starlight as compass. Besides, human activities and disturbances have also caused some insects to get more nocturnal. Climate change has also made some parts of the world drier than before, which may make some insects active at night, preferring to move at dawn and sunset for their activities.

10.5.3 Burrowing under the Sand

The desert is a harsh ecosystem. The high temperature during the day will quickly dehydrate any human body, while the low temperature at night requires a warm shelter. Despite extreme conditions, the desert is a refuge for various life forms. Insects existing in these environments have developed ways to manage temperature fluctuations, scarce water supply and even lack of food sources. For most desert residents, especially small insects such as the dune cricket, they safely live underground until the sun falls down. With nocturnal activities and nest or burrow lifestyles, desert crickets excavate holes in remote tunnels– every insect for itself – within the sand, where they spend all day.

Figure 10.5: Desert insect burrowed under the sand.

Source: https://beetlesinthebush.com/2009/05/11/now-you-see-me/

This is not just for protection against predators. Research has revealed that in the desert, moisture supply of fog and precipitation is equivalent to or even exceeds the amount of rainfall in the region. The tunnel developed from the dunes is designed to allow water droplets stream inside. Not only adults, but even nymph larvae form isolated channels in the sand. If they are close to the water, few insects choose to dig soil near the embankment. Different channels for living for various purposes, such as mating or spawning. The design and overall configuration of the tunnel will guarantee its durability (Weber, 2003).

Burrows are long, oblique and can sometimes be up to 100 centimeters deep. The tunnel length is 40 to 60 centimeters, with a width of 5 up to 5.5 cm. Depth can be 20 to 60 centimeters, some even go down to 100 cm. Typically, the depth is determined through the water table while the drilling is stopped when you notice the presence of water. River species sandstone crickets make their pits at least one meter off a water resource.

In the desert, researchers found that in the summer and winter months, the depth of insect burrows was found to be different, and largely depends on the temperature and comparative moisture of the sand. Burrowing a 100 cm deep channel in quicksand is a laborious effort. For a long time, it was always believed that the open feet are a tool for digging holes in the dunes. In 1953 it was the entomologist F. Carpentier, and then Narain Khattar in the year 1972, who recognized the use of large mandibles insects use to excavate in the sand. Whereas the robust and tough mandible quickly excavates burrows, the exhumed sand is quickly pulled out by the insect's rear feet (Weber, 2003).

The expansion of the posterior tibia prevents the sand from sinking back into the channel. The excavated sand gathers in a small mound, which is used to form a plug that blocks the tunnel entrance. After the insects entered, the tunnel gets closed and stays that way. Once it comes out, it will never go back. A new tunnel can be rebuilt once more. Therefore, the context of the sand plug is a clue for detecting the occurrence of insects. Meanwhile, fresh sand plugs show that the insect is inside.

The top of the sand mass and the breadth of the tunnel suggest if the bug is a grownup or nymph. The backside of the tunnel normally is wider and the bugs live facing with the entrance of the tunnel. Whereas young grownup crickets venture deep searching for water and building closed rather deep tunnels, open tunnels are typically constructed on river banks and filled by older and less energetic insects for which they function as traps for seizing prey.

In the desert environment, the three factors that affect the shape and ecological physiology of its residents are namely high temperature, drought and hostile environment. Nocturnal animals should stay in the retreat and caves every day, and should avoid all extreme conditions of these parameters. However, they still cannot escape completely, because the desert may still be very hot at night, or it may be very cold. It may be very dry, or it may be flooded, and there are never enemies. There is no sense of security in the dark, because the light from the stars and the moon in the clear desert sky is so bright that adaptive colors are very important at all moments (Brown, 2013).

In many insect species that adapt to the desert, especially Coleoptera, the retraction of wings and the lack of flight reflect both its biological and physical features. In arid areas, like trogid beetles, the selective benefit of flight-based scavenging may be offset by the shortcomings of water loss related with this activity that requires metabolism. Whatever the reason for being unable to fly, this condition must make insects more susceptible to enemy attacks than if they could fly away. Therefore, the ability to self-secrete in relatively safe caves increases the value to wingless desert insects. During idle periods, many desert insects hide underneath caves, rock cracks and soil crevices.

Although several Australian desert termite types make up massive mounds, in other regions, desert ants and termites are ground-dwelling. The improved microclimate setting of the cave is outlined, besides temperature regulations may be easily realized by entering the cave. Even though desert woodlouse such as *Hemilepistus reaumuri* are fervent burrowers, anti-social desert insects are often facultative burrowers. For example, most Cerambycidae dig soft sand. At the same time, desert termites inhabit the desert caves of other animals, or hiding in the presence of rocks and stones. The forms of notum and seta insects are characteristic of the larvae of powdery ciliates, however the implication of their adaptation is not yet clear. In desert insects, adaptation measures specifically for digging holes are not obvious (Brown, 2013).

Digging is mainly done through the legs; pedipalps are mainly used for support. The bottom of the cave remains smooth, whereas the edges and ceiling are solidly filled by manipulating the tail. The digging habits of desert insects are closely related to their location of taxonomy, but not the same. For example, as scientists point out, some insect family members with slender claws avoid digging holes. In its place, they occupy debris under rocks and ground fragments. The exceptions are insects which usually live in burrows dug deep underground. In disparity, some species have huge claws and commonly inhabit deep caves that were dug by themselves or excavated.

10.5.4 Rolling Away and Evasion

For animals in danger, it is sometimes best to use the best defense strategy that optimizes costs and benefits. Experiments have shown that when some desert ants are disturbed, they can't move. However, once interference is detected, but under the specific context of the particulate, they will turn to an active escape strategy and then roll away. It's a context-based behavior that is only adopted when appropriate (Weber, 2003).

During the rolling process, prey insects use their antennae and rear legs to become spherical to obtain active motion along a solid trajectory. The advancement of effectual defense systems is among the characteristics behind the accomplishment of desert insects. They are also equipped with various mechanical and substance weapons and behavioral strategies that can make them evade danger. Some insect types that form clusters or have a higher degree of social structure may depend on both personal tactics and collective responses to threats facilitated by alert communication.

In any event, the best defenses are not always able to resist attacks and / or counterattacks. Sometimes, based on ecological and morphological limitations, the best strategy is predator avoidance. Desert insects provide outstanding examples of social and personal defense strategies. Under certain ecological conditions, certain species have evolved their morphological functions and behavioral adaptability, to suddenly get out of the threat.

Like *Harpegnathos saltator* and other insects, jumping is among the most recognizable systems to escape the fastest. This can also be accomplished in unconventional ways, such as in the trap jaw ants, biting their mandible on a solid surface to make it eject from the predator. The Malagasy cliff resident ant *Malagidris sofina* took another special defensive jump action. Their workers fall close to the intruder while dropping off the surface of a cliff, thus protecting their nest from the intruder (Weber, 2003).

The desert insects living on trees adopted another amazing innovation where they can quickly avoid the enemy. These insects fall from the tree and slide along the spiral path, so that they fall safely on the tree trunk again. A similar unique defensive behavior captured by ants is that in the presence of a possible danger, based on a special escape strategy, it curls into a ball and rolls away.

It is difficult to find colonies since they are inconspicuous and occur in rather heterogeneous environments, between rocks, on branches on the ground and in decaying wood, with or without top soil. Workers are very elusive, and they are rarely watched in the wild as small, dark, slowly moving, sweeping and predatory insects. When disturbed, they will freeze or coil the body and become a defense strategy, sometimes called thanatosis, a strategy that occurs in many desert insects.

The duration of the defensive spherical position also varies. When the movement is completed, the ant will maintain the assumed spherical position during rolling for some few seconds and then resume walking. However, if the event that caused the loss of contact with the substrate is more severe, the duration of the location differs depending on the cause of the drop (Weber, 2003).

Methods of defensive rolling: Rolling escape is not a clear method of defense, so the special movement of the desert ants mentioned here is a rather special case. As a matter of fact, to date, just a few animals have reportedly adopted it. The greatest studied group includes the desert animals, which roll along flat sand dunes, moth caterpillars also form wheels that roll backwards to escape interference, and some choose to flip their bodies when wading without any water on wet sand.

Recently, large insects have also been described as being bent into a ball to transport them to a safer spot. There are various bodily and physical limitations to rolling movement, and they are general ways of movement in animals. Nevertheless, if the agent can turn itself into a wheel or a spherical object instead of using the body part as a wheel, they can roll down the slope, jump and bounce to overcome the barrier on their path, this way the danger may not be so severe (Weber, 2003).

This is precisely what some desert ants do in heterogeneous environments (like leaf litters). It's been confirmed that for rolling ants, pebbles and leaves provide a more suitable substrate as compared to the earth, and can reach a farther distance, and then disappear underground beneath materials or plants, or resume walking for escape reasons. Pebbles may provide greater friction and possible obstacles for smaller rolling objects, whereas stones and leaves provide tougher, smoother and curled surfaces, ensuring springboard effects and making rolling ants work like bouncing balls.

In any case, the ground may provide an effective hiding prospect for escaped or frozen ants. They are small and dark, and may be very secretive for vertebrates and arthropods looking for prey in front of their eyes. For other hidden ants, there are similar reports of using slow motions or fixed motions. When the potential danger is imminent, the locomotion strategy together with morphological factors may be used for camouflage.

The rolling motion of insects is quite rapid and needs a set of adjustments, like using the antennae as an arm to finish the body transformation and adopt an effective spherical position to achieve a stable path. From the instant the distress incentive was provided, everything happened in roughly half a second. Like other rolling invertebrates, these rolling prey insects can achieve linear speeds up to 40 cm / s under experimental conditions. This is much better than walking at a speed of 0.5 cm per second in the same research, or by regular movement of other arthropods (such as possible predators) (Weber, 2003).

10.5.5 Adaptations

The rapid extension of back legs will give a kick of sensation when starting to roll, which may help these performances. Notably, scientists have revealed that the escaping strategy through rolling of desert insects depends on the specific situation, and is only actively adopted after reaching a conducive slope threshold (>10 degrees) or due to robust physical interference that makes it lose contact with the substrate (Brown, 2013).

It has also been confirmed that if the reason for the disturbance is an abrupt and unpredicted mechanical factor, the ants normally stay longer after curling, instead of actively escaping by rolling. This may mirror the defense strategy embraced by insects. As a matter of fact, rolling across the gradient after being aware of the possible danger is just the first stage in prey avoidance approach, quickly escaping to find a safe place being the second step.

This is consistent with other environment-related strategies reported in various phases of insect life, including a collective response to dangerous conditions. The rolling action of ants is a naturally appropriate and realistic evasion approach. The effectiveness of this strategy was evaluated as a defensive approach against the enemy for assessing its adaptive value. Rolling provides the insects a great chance to safely escape the assaults of other opponents, thereby greatly reducing casualties (Brown, 2013).

Questions are there about the morphological function and behavioral adaptability that make this rolling conceivable. For instance, it can be speculated that insects have evolved some special forms and maneuverability of bodily parts to create an operational rolling matter, and the unique design of neuro-sensory and motor apparatus, making these context-dependent manifestations possible in jumping and trap-jaw insects. These features are worthy of further study because these ants symbolize a new research prototype that can be applied in the biomechanics and mechanization of rolling items. Finally, the unresolved issue remains to make this behavior unique or barely limit its range among the diverse insect species defined thus far, and caused by strolling and other pre-existing activities (such as pupal-shrinking poses) The evolutionary model assumed allows for easy escaping same as a rolling stone (Brown, 2013).

10.5.6 Feigning Death

This defensive strategy is scientifically called Thanaaosis. Carnivores quickly lose attention in dead prey, therefore, insects that apply extermination strategies can usually escape without harm. The act of pretending to die usually resembles a stop, fall, and roll because the threatened insects will let go of any substrates they touch and fall on the ground. Then they become still, delaying for the hunter to leave. Bugs that avoid predation by feigning death include some caterpillars, ladybugs ,beetles, weevils, giant water bugs and robber fruit flies.

Figure 10.6: The beetles of genus Cryptoglossa are called by the collective name of death-pretending beetles.

Source: http://cincinnatizoo.org/news-releases/blue-death-feigning-beetle-emerges-in-time-to-celebrate-world-of-the-insects-40th-anniversary-at-the-cincinnati-zoo/

When seeking to collect dead insects, it is usually easy to put a collection pot or dish under the twig or substrate where the insect originated. Thanatosis, also called tonic immobility or (TI), is an underreported but intriguing anti-predator approach that is adopted by various insect predators in the late stages of the predator sequence and often after physical connection with the predator.

TI is believed to suppress further attacks by carnivores and lessen the perceived need for carnivores to further subdue their prey. This behavior may exist in more taxa as compared to what's currently described, however for some practical and possible ethical reasons, the exact taxonomic distribution is uncertain even in groups that have been thoroughly studied. Researchers are still studying the form, role, evolution and ecological costs, plus advantages of TI. Further contemplating the possible evolutionary impact of some predator forms on the growth of predator strategies, and other non-defensive cases where TI may occur (Tarbox, 2007).

A common belief is that it's necessary to better understand TI in scientific literature and to outline possibly profitable research approaches. The future of using technology in the desert should bring beneficial development to this research field. Anti-predatory defense is essential to many facets of behavioral ecology. Despite its extensive taxonomy and ecology, Thanatosis, has for long been an unknown defense (Tarbox, 2007).

Understanding its significance: Predation is essential in the life of desert animals. It affects key aspects of adaptability, such as eating, reproduction, and even mortality. In reaction to the risk of predation or the adaptation costs of individuals associated with this predation, still natural selection has led to various morphological, biological, chemical, and behavioral defense adaptation changes in prey species. Few insects are immune to predatory risks during at least a given stage of their individual development. Therefore, behavioral ecologists are very concerned about the many and occasionally complex adaptations involved in relations between predators and prey (Patent, 2003).

The interaction between the predator and the prey can usually be divided into several stages, starting from two creatures approaching and starting to detect, identify, reduce separation, contact, and subduing the other. In order to avoid being attacked by predators, various mechanisms have evolved, some of which need to be deployed at specific stages of the attack sequence to be effectual. A captivating and well-recognized defense that prey can show a later state in the system of predator events is tonic immobility, hereafter referred to as TI.

TI has been debated in research literature for well over a century. However, until recently, many studies were observatory and qualitative. Being a defense method, TI can be considered as the "last resort" option, not simply after the predator has detected the prey, but most so after physical contact between predator and the insect prey. This makes it very different from the static state used to lessen the danger of predators being discovered or tracked, because this "freezing" happens earlier in the system of predatory attacks. In other words, researchers believe that enhancing the stillness or camouflage tactics provide a different but complementary technique from TI.

In addition, certain animals (example, pill bugs) adopt a curly posture, preventing predators from accessing fragile body parts. This too is an independent process from TI, which doesn't depend on making prey physically vulnerable to predators. TI usually occurs when the animal is restrained by the body. It involves the prey in a relatively fixed state. Even if the physical restraint is released, this state may last for a few seconds to several hours. In some vertebrates, this behavior may result in decreased breathing rate, bradycardia, protruding tongue, and wide-open eyes, reminiscent of deceased individuals of the insect species. Therefore, TI is also called death dress up or animal hypnosis (Patent, 2003).

At the same time, (TI) is a motionless posture taken by the prey without learning, which is caused by physical interaction or very risky non-injury caused by the attacker (or other adversary). This posture does not diminish the predator's capacity to find or identify prey, nor does it diminish the physical susceptibility of the prey insect when attacked. Even after the predator has released, the state of motion inhibition will still remain for some period of time. When in this form, the prey's reactivity to external stimuli decreases (even though the environment may still be monitored).

In absence of death or injury during TI, the prey may recover from its unique physiological state after TI emerges. Note that this description has nothing to do with TI's function. Though TI is usually considered to have anti-predatory benefits, it's believed that this feature is not essential for a perfect and operative definition before the predator is interrupted. Indeed, the definition does not assume a predatory reaction to TI. It also makes no assumptions about the fundamental drivers of any predacious response.

It is also for this particular reason that scientist prefer calling this activity tonic immobility (TI), rather than thanatosis which is from the Greek word meaning "deceased" or death pretense. Regarding function, it is largely assumed that following an active resistance to attack that ends in failure, and as the probability of escape is reduced, TI may be applied to suppress additional attacks by predators and diminish the supposed need for the predator to further subdue the prey. Like most biological adaptations, the nature of TI varies greatly between species and individual insects (Patent, 2003).

For example, physical contact isn't a necessary condition for triggering animal behavior. It is reported that various animals will enter a state of rest when they feel the threat of being attacked without the opportunity to escape. This phenomenon is extensive in taxonomy, however it's among the more difficult study topics in the area of interaction between natural enemies, especially in vertebrates. Since it appears late in the predator sequence, there are moral challenges in conducting research in the laboratory, plus logistical challenges in real-life environments, because naturally happening prey capture is irregular in space and time, and is often restrained by the presence of human observers who may cause fear on the study insects.

Despite being cited as early as the 19th century, particularly in the works by Darwin (1885), TI hasn't been a subject of in-depth research. For the majority of the entire 1900s, it was still a relatively vague and anti-predatory behavior. Nevertheless, in the last 20 years, there have been more purpose-oriented studies on this phenomenon, so it is time for researchers to review this remarkable defense strategy further (Patent, 2003).

Physiological mechanisms of tonic immobility: Controlling TI may involve many different physiological mechanisms. In vertebrates, respiratory rate decreases during TI, bradycardia, salivation, excrement, and urination are widespread, and are matching with mediation through the parasympathetic nervous structure. The effect of serotonin may be complicated, depending on the type, dose and action site, weakening or enhancing TI. In addition, it was also found that the injection of the stress compound corticotropin-releasing factor or (CRF) into the amygdala of the brain prolongs the extent of TI, which is an important area for the regulation of emotional conditions.

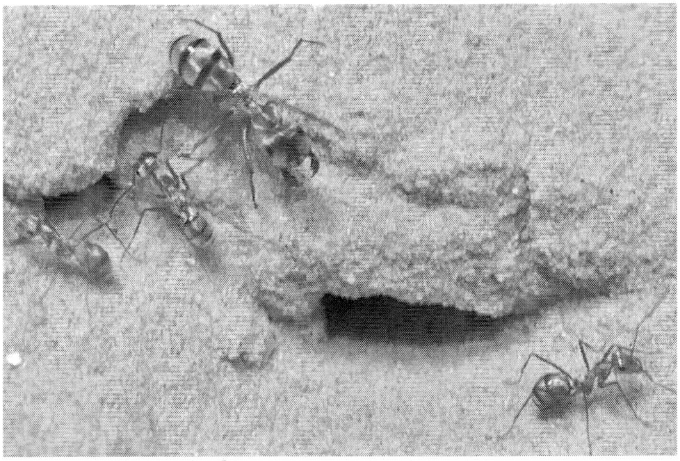

Figure 10.7: Desert ants struggling to locate a moth hidden in the sand.

Source: https://www.sciencefocus.com/news/disgusting-tasting-moths-dont-bother-trying-to-escape-predators/

It was also found that neurochemicals that impact the state of motivation also impact TI. For instance, in some desert insects, dopamine receptor antagonists surge TI duration, whereas dopamine agonists diminish TI duration. In laboratories, TI can be prompted by processing and manually limiting animal movement. This has caused some all-inclusive observations of a variety of desert species, including insects (Auch, 2003).

The pressure of experiencing TI affects many blood chemistry parameters and reduces the breathing efficiency. Nevertheless, the researchers also discovered that a sequence of compensation mechanisms seem to offset these effects. Due to the reduction in the short-term ventilation efficacy caused by TI, the ventilation rate is significantly increased compared to control animals, which increases the gas exchange capacity. In addition, other researchers speculate that the bodily adaptation circulatory system and the adaptability related to stress response were maintained during TI. These adaptation measures include vasodilation of blood vessels, which will help further promote the reduction of carbon dioxide concentration.

The makeup of TI has further been fully studied in desert insects. Lately, studies on beetles have begun to reveal midbrain circuits related to the controlling of TI in bugs. It has been reported that TI induction impacts the circulation and density of c-Fos markers (as markers of neural activity) in particular parts of the bug's brain. They discovered that inducing TI responses by inhibiting test insects can increase the expression of c-Fos in the insect brain complex ICo / GCt, which suggests that ICo neurons might be essential for TI expression.

Interestingly, scientists believe that the practical similarity of insect and bird midbrain circuits in expressing fear and defensive behavior could mean that they're basically conserved in amniotic creatures, which is inevitable in reacting to threatening stimuli. During TI, it decreases strongly, whereas slow flexor and extensor neurons fire in pitch to maintain a stiff pose. For example, these reactions are all present in crickets, but heart rates during TI in this insect taxon seems to be twice that of an immobile state (Auch, 2003).

The researchers described TI's mechanism in crickets and divided the procedure into two different phases: initiation and maintenance. Physical limitations induce TI, which causes excitatory motor neuron burst activity to drive a strong resistance reflex. In turn, this generates isometric tension in the flexor muscles of the insect, which triggers the maintenance response. The characteristic of the maintenance response is that the rapid motor neurons quickly stop their activities, and only the slow flexor body neurons are activated.

In some invertebrates, like the Tribolium beetle as well as *Larinoides cornutus* spider, neurochemicals (like octopamine or dopamine) have been discovered to be key in influencing various TI factors, particularly in Tribolium beetles. The impact of biogenic factors on TI have been broadly studied. Among the flour beetle (*Tribolium confusum*) as well as red flour beetle (*T. castaneum*), lowered brain dopamine levels are seen in strains with lower TI frequency and smaller TI duration, they have greater frequency and extended TI periods (L strain).

In addition, it has been found that the administration of caffeine reduces the TI period in L-strains *T. confusum* as well as *T. castaneum*, besides since caffeine is recognized to stimulate dopamine levels in these species, through the dopamine receptor which affect tonic immobility. As additional taxa are researched for TI behaviors in both vertebrates and invertebrates, scientists expect that more physiological adaptations related to anti-predator strategies will be discovered (Patent, 2003).

Harsh desert conditions that insects often face: Global warming is escalating the occurrence of droughts that dry out wet patches. Higher temperatures may cause increased wildfires, which change the desert landscape by removing slow-maturing trees and bushes and switching them with fast-growing grass. Grazing livestock also destroys various desert plants and creatures that insects depend on for their survival. Not to forget human activities, such as off-road desert driving, oil and gas generation, and nuclear waste clearance, which, if carried out irresponsibly, may cause irreversible destruction to desert environments.

10.6 WILDFIRES

In the desert, we have allowed livestock to ravage native grass, and permitted non-native grass and other exotic plants to infiltrate bare land. The incidence of fires has been growing, causing negative impacts on native faunas and plants. Native grasses are scattered throughout the soil and usually limit the dispersion of wildfires in desert environments. In contrast, intrusive species (for example, the red Brom plant and African buffalo) spread like carpets, stimulating wildfire expansion and typically eliminating native desert shrubs. In such an environment invasive species benefit while local plants suffer (Brown, 2013).

Fuel and fire conditions in the desert have changed due to previous land managing practices, biological invasions, and contemporary extreme weather factors and climate changes. These variations have brought increased fire risks to native and regional insect communities, affecting their health as it related to wilderness biodiversity. In the desert, fire is not only a key natural ecological process, but also sometimes endangers other wildlife and desert resources. It's well recognized that active fire suppression in the past has caused irregular fuel accumulation and excess fire risks.

Science is needed for answering these questions. How does fire impact sensitive species, such as endangered insects? What role does fire and other factors (including climate) play in promoting plant incursion and causing changes in local vegetation coverage? Historically, in the deserts of the Western Hemisphere, fires rarely occur, and are generally considered to be a source of pressure on ecosystems, posing a threat to native insect communities. The invasion of plants and the change of climate have essentially changed the behavior of fuel and fire (Brown, 2013).

The method used has transformed the fire situation from historical norms. Even though these changes are common in desert areas, the magnitude and severity of the changes vary greatly between the main vegetation types. Scientific knowledge is required to understand where, when and how fires can be triggered and what measures may be taken to alleviate these threats.

When a wildfire is finally brought under control, the environmental damage on the fire's wake reflects the heavy wildlife loss and plant habitats, occasionally including endangered species, plus it leaves exposed land vulnerable to intense erosion. Places that are personally important sometimes-such as, the magnificent Santa Clara Canyon within the Santa Clara Pueblo Reserve in north New Mexico-have been damaged by wildfires that can't be easily repaired in a lifetime. Some insects that lived in the area were completely wiped out by this wildfire disaster (Brown, 2013).

10.7 CLIMATE CHANGE IMPACT TO DESERT INSECTS

Changes in the climate of the entire earth have led to variations in the environment. Among them is increase in the extent of arid land that covers the globe's surface. As people find more and more desert areas where they are located, and the annual rainfall is less than 50 cm, it becomes more and more significant to comprehend the environmental stability challenges facing the desert environment, and human activities have exacerbated many challenges (Brown, 2013).

In addition, although deserts occur in hot and cold environments, they still are characterized by limited yearly precipitation. Since tree and grassroots may capture any rainfall, they are usually not widespread across the desert, so the desert area has almost no water, therefore making the water very rare. Insects that come to stay in the desert utilize this resource and change the ground cover through removing plants when dispersing. The loss of desert biodiversity leaves less water inside the soil and may cause soil erosion, which will further hinder plant rooting.

However, the lack of water is not the only factor causing soil degradation. Deforestation, over-cultivation of crops, and excessive grazing of livestock in the Gobi Desert in China all led to the desertification of the land and limited the quality of existent desert soil through depriving the soil of nutrients that sustain insect life. However, liable irrigation and farming approaches have been revealed to enhance nutrient content (plus water retention) in desert soils.

Storms can lift sand into the air, which is another means of depriving important soil nutrients present in the desert. In addition to affecting soil quality, dust storms may make breathing hard and even block sunlight required for normal plant growth. In cities, neighboring deserts, such as Tucson in Arizona State, fungal spores have germinated during urban development and these fungal spores have invaded the lung tissue of insects dwelling in urban regions, damaging their health and populations. Variations in desert habitats might make it hard for native insect species to survive (Auch, 2003).

Additionally, established insects might be threatened by organisms that have just landed in the deserts that are more appropriate for that climate and ecosystem. These species might migrate to deserts naturally, or they might be brought even involuntarily, by individuals who travel there. Whichever way, they might compete with the given species for possible resources, therefore presenting another risk to the subtle balance of the desert environment.

10.8 DESERT STORMS AND FLASH FLOODS IMPACTS ON INSECTS

Sandstorms are severe storms that often occur in deserts. In the Middle East, sandstorms may occur and stay for even 3 months. Although these winds will not kill humans, they often cause environmental disasters due to the blinding outcome of sand. Insects may be caught up in these storms that are destructive, either causing mass extermination in the local desert or transporting them far from their natural habitat, where they may end up perishing due to lack of proper adaptation (Auch, 2003).

Sandstorm conditions are furthermore perfect for causing heavy rains, whereby flash floods become a threat to insect populations. The sand in the desert does not absorb water quickly, therefore, heavy rain can quickly cause flooding without prior warning. Dry River channels, trenches and lake beds may fill up rapidly, besides the water potential can be violent-at times forming water walls between 10-to-30ft high. It is worth noting that more insects drown in deserts from flash floods than drought or thirst.

Due to the risk of flash floods, even if it does not seem like rain, animals must not rest or sleep within trenches or dry creeks. Desert deluges occur quickly without any form of warning, plus may uproot trees or move boulders where insects stay. In the case of flash floods, some insects will reach higher grounds as quickly as possible and avoid perching around rocks or trees. They'd better be 30 to 40ft above the lowest deluge point (Wehner, 2019).

These storms may be rough, but typically short-lived. The best bet for desert insects surviving flash floods is keeping their eyes out and anticipating its arrival. Many insects that perish in these floods usually are captured off-guard. Being alert for thunder as well as lightening in the local area can also help insects survive. If they suspect a storm is impending, they can get simply get to higher ground and then wait it out.

To summarize, favorable conditions for the survival of desert insects vary depending on the species. Many have developed adaptations that are suitable for their environment, such as nocturnal movement to conserve water which is a scarce resource in the desert (Wehner, 2019).

References

1. Allaby, M. (2009). Grasslands. New York: Infobase Publishing, 1-81.
2. Aloian, M. (2018). Where do insects live?: Read Along or Enhanced eBook. New York: Triangle Interactive, Inc, pp27.
3. Auch, A. (2003). Desert Animals. New York: Capstone,.
4. Bobbie Kalman & John Crossingham. (2006). Insect Homes. Canada: Crabtree Publishing Company, pp32.
5. Brown, G. W. (2013). Desert Biology: Special Topics on the Physical and Biological Aspects of Arid Regions. New York: Elsevier.
6. Campos-Herrera, R., El-Borai, F. E., & Duncan, L. W. (2015). Modifying soil to enhance biological control of belowground dwelling insects in citrus groves under organic agriculture in Florida. . Biological Control,, 84, 53-63.
7. Collins, N. (2012). The Conservation of Insects and Their Habitats. Massachusetts, United States: Academic Press, 33-71.
8. Collins, N. (2012). The Conservation of Insects and Their Habitats. New York: Academic Press, 33-71.
9. David Hagstrum & Bhadriraju Subramanyam. (2016). Stored-Product Insect Resource. New York: Elsevier, pp518.
10. David W. Hagstrum, Thomas W. Phillips, & Gerrit Cuperus. (2012). Stored Product Protection. Kansas, S156 ISBN 978-0-9855003-0-6: Kansas State University.
11. Dhang, P. (2014). Urban Insect Pests: Sustainable Management Strategies. Wallingford, United Kingdom: CABI, 1-99.

12. Einspruch, A. (2013). Rain Forest Habitats. New York: The Rosen Publishing Group, pp32.
13. Fleisher, P. (2013). Grassland Food Webs in Action. Minnesota, United States: Lerner Publications, 5-34.
14. G. Achuthan Nair, John C. Morse & Stephen A. Marshall. (2015). Aquatic insects and their societal benefits and risks. Journal of Entomology and Zoology Studies , 3(3): 171-177.
15. Gerhard Eisenbeis & Wilfried Wichard. (2012). Atlas on the Biology of Soil Arthropods. Berlin: Springer Science & Business Media, 1-90.
16. Gill, H. K., McSorley, R., & Branham, M. (2011). Effect of organic mulches on soil surface insects and other arthropods. Florida Entomologist,, 226-232.
17. Hagstrum, D. (2016). Fundamentals of Stored-Product Entomology. New York: Elsevier, pp323.
18. Heckman, C. W. (2008). Encyclopedia of South American Aquatic Insects: Odonata - Zygoptera: Illustrated Keys to Known Families, Genera, and Species in South America. Berlin: Springer Science & Business Media, 10-102.
19. Holt, G. (2012). The Mystery of the Secret Caves. Bloomington: Xlibris Corporation, pp.105.
20. Hopkins, T. (2009). Insect Homes. New York: The Rosen Publishing Group, pp24.
21. Ingo S. Wehrtmann & Jorge Cortés. (2008). Marine Biodiversity of Costa Rica, Central America. Berlin: Springer Science & Business Media, 1-47.
22. Ivan Hiltpold & Michael Rostás. (2017). Grassland-Invertebrate Interactions: Plant Productivity, Resilience and Community Dynamics. Cape town: Frontiers Media SA, 1-105.
23. Jong, G. D. (2018). Aquatic Insects in the Vicinity of the Black Hills, South Dakota and Wyoming. New York: Lulu.com, 25-85.
24. Judicaël Obame-Nkoghe, Eric-Maurice Leroy, and Christophe Paupy. (2017). Diversity and role of cave-dwelling hematophagous insects in pathogen transmission in the Afrotropical region. Emerging Microbes & Infections, 6, e20; doi:10.1038/emi.
25. Kalman, B. (2009). The ABCs of Insects. Canada: Crabtree Publishing

Company, pp32.

26. Kleber Del-Claro & Rhainer Guillermo. (2019). Aquatic Insects: Behavior and Ecology. Berlin: Springer, 1-95.
27. Klotz, J. H. (2010). Urban Pest Management of Ants in UCANR Publications. UCANR Publications: UCANR Publications, pp.72.
28. Leather, S. R. (2008). Insect Sampling in Forest Ecosystems. Florida: John Wiley & Sons, 1405140291, 9781405140294, pp340.
29. Mills, J. E. (2009). The Creation of Caves. New York: The Rosen Publishing Group, pp.64.
30. New, T. R. (2011). n Considerable Variety': Introducing the Diversity of Australia's Insects. Berlin: Springer Science & Business Media, 1-93.
31. Paine, T. D. (2007). Invasive Forest Insects, Introduced Forest Trees, and Altered Ecosystems: Ecological Pest Management in Global Forests of a Changing World. Berlin: Springer Science & Business Media, pp189.
32. Patent, D. H. (2003). Life in a Desert: The Sonoran Desert. Minneapolis : Twenty-First Century Books, PP72.
33. R. N. Gibson, R. J. A. Atkinson, & J. D. M. Gordon. (2004). Oceanography and Marine Biology: An Annual Review Volume 42, Volume 42. FLorida: CRC Press, 1-95.
34. Robinson, W. H. (2005). Urban Insects and Arachnids: A Handbook of Urban Entomology. London: Cambridge University Press, 6-74.
35. Rufer, M. (2015). A Children's Guide to Aquatic Insects. Minnesota: University of Minnesota.
36. Szujecki, A. (2012). Ecology Of Forest Insects. Berlin: Springer Science & Business Media, 1-78.
37. Tarbox, A. D. (2007). A Desert Food Chain. Minnesota: The Creative Company, pp43.
38. Weber, R. (2003). Tricky Insects: And Other Fun Creatures. Minnesota : Capstone, 1-91.
39. Wehner, R. (2019). Desert Navigator: The Journey of an Ant. London: Harvard University Press, 1-91.
40. Weissteiner, S., & Schütz, S. (2006). Are different volatile pattern influencing host plant choice of belowground living insects. . Mitt Dtsch Ges Allg Angew Ent, 15, 51-55.

INDEX

F

G

H

I

Q

R

S